Feminist geographies

Explorations in diversity and difference

WOMEN AND GEOGRAPHY STUDY GROUP

of the Royal Geographical Society with the Institute of British Geographers

LONGMAN

Addison Wesley Longman Limited
Edinburgh Gate,
Harlow
Essex CM20 2JE
England
and Associated Companies throughout the World

First published 1997

ISBN 0 582 24636 9

British Library Cataloguing-in-Publication Data
A catalogue record for this book is
available from the British Library.

Library of Congress Cataloguing-in-Publication Data
A catalog entry for this title is
available from the Library of Congress.

Set by 30 in 10/12 Sabon.
Produced by Longman Singapore Publishers (Pte) Ltd.
Printed in Singapore

Contents

List of figures

List of tables

Contributors

Chapter convenors:

NICKY GREGSON	Department of Geography, University of Sheffield
VIVIAN KINNAIRD	School of Environment, University of Sunderland
UMA KOTHARI	Institute for Development Policy and Management, University of Manchester
NINA LAURIE	Department of Geography, University of Newcastle upon Tyne
CLARE MADGE	Department of Geography, University of Leicester
MANDY MORRIS	Department of Geography, Open University
CATHERINE NASH	Department of Geography, St David's University College, Lampeter
PARVATI RAGHURAM	Department of International Studies, Nottingham Trent University
GILLIAN ROSE	Department of Geography, University of Edinburgh
TRACEY SKELTON	Department of International Studies, Nottingham Trent University

Other contributors:

SOPHIE BOWLBY	Department of Geography, University of Reading
JULIA CREAM	Independent researcher
CLAIRE DWYER	Department of Geography, University College London
JO FOORD	Department of Geography, North London University
SARAH HOLLOWAY	Department of Geography, University of Loughborough
AVRIL MADDRELL	Department of Geography, Westminster College, Oxford
SARAH MONK	Department of Geography, Anglia University
SARAH RADCLIFFE	Department of Geography, University of Cambridge
JO ROWLANDS	Department of Geography, University of Durham
FIONA SMITH	Department of Geography, University of Dundee
JANET TOWNSEND	Department of Geography, University of Durham
JENNY WILLIAMS	Wirral Metropolitan College
KATIE WILLIS	Department of Geography, University of Liverpool
LIZ YOUNG	Department of Geography, Staffordshire University

Acknowledgements

This book has been a long time coming and in many senses it is perhaps appropriate to acknowledge here the debts we owe collectively to one another. Others, however, have been supportive in different ways. Our thanks therefore go to Ann Rooke and Noelle Wright for their cartographic input and to Doreen Massey for her support of this project through its various stages.

We are grateful to the following for permission to reproduce copyright material:

Fig. 6.2 from *Flower Abstraction* © ARS, NY and DACS, London 1997 (Georgia O'Keefe, 1924); Fig. 6.3 from *Cottage at Chiddingfold* (Allingham, 1889), image from the Victoria and Albert Museum; Fig. 6.4 from *Pastoral Interludes* (Pollard, 1987); Fig. 6.5 from *Another View* (Pollard, 1993); Fig. 6.6 from, © The Estate of David Wojnarowicz, *I Feel Vague Nausea*, 1990 from *Brush Fires in the Social Landscape*, Aperture, New York, 1994.

Canadian Association of Geographers for extracts from articles by P Gould, L Peake & P Gould in **Canadian Geographer** *38/3* 1994. pp194–200. pp204–6. & pp209–14; Clark University for extracts from an article by S Hanson & G Pratt in **Economic Geography** *64/4* 1988. pp299–318; Elsevier Science for an extract from an article by S Brownhill & S Halford in **Political Geography Quaterly** *9/4* 1990. pp396–414; Guardian News Services Ltd for extracts from articles 'When I was a girl...' in **New Internationalist** (*Coming of Age in Conchali*) February 1995. pp12–14, 'The EPZ economy...' in **New Internationalist** (*The Asian Miracle*) & 'Foot in the door domestics fight servility and the sack' in **New Internationalist** (*Death Without Weeping*) No 254 1994. p29; Institute of British Geographers for an extracts from articles by D R Stoddart & M Domosh in **Transactions of the Institute of British Geographers** *16* 1991. pp484–7 & pp488–90. & an extract from an article by G Valentine in **Area** *21/3* 1989. pp385–90. International Thomson Publishing Services for an extract by M L Pratt in **Imperial Eyes: Travel Writing and Transculturation** pp214–5. pubd. Routledge: London 1992; Zed Books Ltd for a part poem by Ghanshyam 'Shalland & an extract from **Women and the Environment** by A Rodda. 1991.

We have been unable to trace the copyright holder of the poem by Bronwen Wallace and would appreciate any information which would enable us to do so.

CHAPTER ONE

Introduction

GILLIAN ROSE, NICKY GREGSON, JO FOORD,
SOPHIE BOWLBY, CLAIRE DYWER, SARAH HOLLOWAY,
NINA LAURIE, AVRIL MADDRELL AND TRACY SKELTON

1.1   Why and how we wrote this book

In a review article published in 1993, Linda McDowell remarks on the extent to which feminist geography has evolved, both theoretically and empirically, in the decade since its inception. Such reflections, in many ways, underpin our intentions here. Having been involved in producing some of the research which has constituted, shaped and made feminist geography over the past ten or so years, and having invested considerable time and resources in designing and teaching courses which reflect these research interests, many of those active within the Women and Geography Study Group (henceforth WGSG)[1] feel that now is an appropriate time in which to produce a teaching text which does justice to the breadth, diversity, intellectual vibrancy, debate and difference currently to be found in feminist geography. This has been our primary objective in writing this book, and we certainly hope that in the pages which follow we manage to communicate to you something of the excitement which we feel characterises feminist geography in the 1990s.

   As important as conveying the vibrancy of debate in feminist geography, however, is the way in which this book has been written. In 1984, when feminist geography was just getting going, and when the field was charac-

---

[1] The WGSG was established in 1980 as a study group of the Institute of British Geographers (IBG), the then professional organisation of academic geographers. The group's aims since then have been to encourage the study of the geographical implications of gender differentiation in society and geographical research from a feminist perspective; and to encourage and facilitate the exchange of information and ideas with reference to research and teaching in these areas. In 1994 the IBG merged with the Royal Geographical Society (RGS) to become the RGS(IBG). The merger itself attracted considerable opposition from radical geographers, including feminist geographers, many of whom see the RGS as embodying some of the worst aspects of Geography's past and present; for example, its connections with empire and an imperial past, its collusion with the interests of capital, and its patriarchal organisational structure, encapsulated in the labelling of members as fellows. With the merger, the majority of WGSG members became fellows of the RGS(IBG) and are actively trying to change the organisation from within. However, others choose to be associate members of the WGSG, and take no part in the wider RGS(IBG) structures. Amongst both groups numerous individuals are working towards the formation of an alternative professional organisation to represent academic geography.

terised by a relatively small number of individuals, the WGSG produced the text *Geography and Gender*. Aimed at first-year undergraduates, this ground-breaking book was also important for the collective way in which it was written. Rather than being written by one or two named individuals, the book was written by several people working and writing together, as 'a collective', and has as its 'author' the WGSG. Taking their cue from wider feminist politics, the feminist geographers who wrote *Geography and Gender* wished both to challenge the accepted conventions of academic writing (conventions which celebrate and reward the individual, apparently working in glorious isolation) and to acknowledge the genuinely collaborative and supportive ways in which feminist geography was emerging in Britain. And it is in the same collective tradition that this book has been written. Each chapter, therefore, has been written by different groups of named individuals and subsequently revised and redrafted in consultation with a core group of chapter convenors, who have also taken responsibility for the overall coherence of the book. Two reasons in particular have influenced our decision to do things in this way. First, we wanted to echo deliberately the spirit of the first feminist geography text; to reaffirm the importance and possibilities of collective writing and the tradition of working in this way which *Geography and Gender* established. Secondly, and linked to this collective, supportive tradition, we wished to acknowledge in a teaching text the extent to which the teaching of feminist geography remains a collective enterprise. Although located in a number of higher education establishments across Britain, those of us who teach feminist geography regularly share our teaching resources, our experiences and our strategies (including those that work and those that are not so successful!). Moreover, teaching for many of us is a forum in which students are encouraged to engage in critical debate and to formulate their own position(s). As a consequence, therefore, many of us have found that what we have needed as a teaching text is a book which facilitates just this, something which enables students to explore, discuss and debate fully a range of different positions but in a context which is liberated from the guiding voice and guiding trajectory of one teacher and one course. Accordingly, *Feminist Geographies* admits precisely this: multiple versions and multiple visions of feminist geography.

So much for the whys. What of the hows? Any of you who have experience of working or being assessed as part of a group will have some inkling of the difficulties which can be encountered when working collectively.

ACTIVITY

If you've ever worked or been assessed in your undergraduate studies as a group, spend a few moments recapping some of the problems which you encountered in this way of working. Then think about the advantages which such ways of working offer. If you have no experience of such styles of teaching, what potential advantages or disadvantages can you think of?

Predictably, writing this book has involved different positions and consequent disagreements; voices which speak from varying positions of authority[2]; negotiations and compromises; different personal and professional pressures; different degrees of commitment; frustrations over styles of writing and how long people take to accomplish particular tasks, etc. The list may be familiar to you, as may be our means of resolving dispute – democratic consensus, with ultimate decisions resting with the group of chapter convenors. However, it is one of the earliest decisions which this collective took which we wish to dwell further on here, and this is the exclusion of men from the writing collective.

The decision to exclude men from the writing team was not an easy one. It is also a decision which has been challenged publicly (Pinnegar, 1996; Silk, 1995). Reflecting on this decision here, we want to make explicit that feminist practice is not something which most in this collective take to be inherently tied to female experience: most in this collective are not of the opinion that feminism is the exclusive prerogative of women (and see Scott, 1992). But some, however, are. For them, whilst men can sympathise with feminism and be anti-sexist, they cannot be feminists (and see Stanley and Wise, 1993), and part of the reason why men were excluded from the writing collective reflected this argument. Beyond this, though, and more critically, the decision reflected the politics of writing. Along with many feminists, we see writing as a political activity; not only is writing about trying to secure change, but in the act of writing we make political choices about whom we write with, for particular reasons. Many of those involved in the *Feminist Geographies* collective are at the start of their academic careers. Having recently completed (or indeed, in some cases, still completing) their doctoral theses, and frequently working as the one woman on the staff in their department, many of those involved in this book felt the need to discuss and write in a women-only space, away from the hierarchies, assumptions, language and conventions of the male-dominated world of Geography. Moreover, for all of us in this collective – including the minority who have been in academia somewhat longer – writing in a women-only space is not just about seeking a haven but about creating a space of resistance, a space from which to challenge the hierarchies, assumptions, language and conventions of the male-dominated world of Geography. Wanting both a space to call our own and a space from which to contest and resist, the political choice we made was to exclude men from the writing collective, although we should perhaps point out that only two men expressed a definite interest in this project at its outset.

---

[2] Authority in an academic context relates primarily to differences in position both between individuals and between institutions. It is fundamentally about professional status, professional standing and the consequent weight of individual voices to influence and shape decisions. Certain voices then 'count' more than others. In writing this book there have been differences between 'established' and younger voices, between those on permanent and short-term employment contracts, and between those from 'new' and 'old' universities.

Table 1.1 details the authorial balance by gender of papers published in the journal *Area* over the period 1993–94. What patterns can you identify here? Spend some time thinking about issues of writing strategies and writing spaces. Why do you think that the published output of human geography is dominated by single (male) authored texts? How do you think this might connect with our decision to exclude men from this writing collective?

Table 1.1 Gender composition of articles submitted to *Area* in 1993 and 1994

	1994		1995	
	Number	**Percentage**	**Number**	**Percentage**
Single author				
female	6	8	14	23
male	51	68	29	48
Joint authors				
female	3	4	0	0
male	12	16	13	22
mixed	3	4	4	7

Source: *Area*, 27(2), 99.

1.2   What is feminism? What is feminist geography?

Right now there are probably two questions which are likely to be uppermost in your mind as you read this book. These are 'what is feminism?' and 'what is feminist geography?' Rather than provide you with simple definitions here, however, we want to get you to start answering these questions for yourselves. This, you might think, is to duck the issue on our part. 'What', we can hear you say, 'are teachers for, except to provide students with the information which they need to know?!' Well, though we agree in part with this position, where we part company with many students is in how we achieve this. Thus, rather than providing you with 'the right answer', we prefer to teach by helping students to construct their own answers, positions and responses. This is particularly important with respect to questions of feminism and feminist geography. Indeed, there are varying and conflicting interpretations and understandings of feminism, just as there are differences between us as authors in what we interpret the term feminist geography to mean. Feminism, therefore, is a contested term, one whose meaning is disputed. And, for many of us, it is now more appropriate to use the term feminisms rather than feminism. The following activity is designed to get you started on thinking about what you currently understand by the term feminism.

Listed below are several definitions of feminism. Spend some time either thinking about or discussing these definitions. Then, identify the two or three definitions which most closely approximate to your current understanding of feminism. If you want to add other definitions to the list please do.

- The study of women
- The assertion that women are more important than men
- A focal point which enables women to celebrate being female
- The prioritisation of gender as an analytical category
- The campaign for equal rights
- A movement which unites all women
- Working for women
- A political project which aims to improve the representation of women in social, economic and political spheres
- A broad heading incorporating many and various concerns and activities of different women
- 'Women's Lib'
- A 'yesterday's movement'

The majority of these definitions are taken from various strands of the feminist move-ment, both political and academic. Some, however, are non-feminist definitions; some relate to earlier times; others are more contemporary. As you worked through this exer-cise did you find it easy to dismiss certain definitions? Why? Did you find that some definitions seemed to be subsumable within others? Did you at any time feel that just one definition of feminism would suffice? If not, what does this suggest to you about feminism? You will need to keep a note of your answers here, and as you work through this book consider whether your position alters. In the conclusion to the book we will ask you to think again about this activity.

Much as we are reluctant – for academic and pedagogic reasons – to pro-vide you with a straightforward, simple definition of feminism, so we are reluctant to define feminist geography in such a way too. Indeed, just as there are multiple versions of feminism – feminisms rather than feminism – so too are there of feminist geography. And since this book itself is an exploration of these various feminist geographies, then in many senses we want to leave the task of unravelling these various feminist geographies to you. One point, however, on which we are all agreed, is that notwithstanding the existence of feminist geographies, the practice of feminist geography is one which contests and challenges the frequent taken-for-grantedness of the content and concepts which sit at the heart of the discipline of geography, and their assumptions. Accordingly, and so that you can understand both the resistant and reconstit-utive nature of feminist geography, it is imperative that we say something here by way of an introduction to some of the core geographical concepts and their conceptualisation outside feminist traditions. We focus on space, place and environment – the three concepts which underpin later chapters of this book and which have concerned feminist geographers in much of their work.

Space

Perhaps the most obvious way of thinking about space is as a physical reality. We know that objects exist in space; that they have a unique location (latitude and longitude); and that to overcome the separation between objects, to move between them, or to bring them together, we must exert energy. Space thought of in this way seems an empirical fact, indisputable and fixed. This conception of space is often referred to as absolute space. Indeed, locating objects in absolute space was a preoccupation of early explorers and mapmakers and was of great economic and social significance. Knowing once and for all where a place was on the surface of the globe was important to the development of trade and to the political domination of one country by another.

In contrast to absolute views of space is the idea of relative space. Here space is not seen in terms of the absolute location of objects. Rather it is the relative placement of objects to other objects which matters. In the 1960s many British and North American geographers tried to develop this approach to space within geography, studying the similarities between the spatial patterns of, say, rivers and transport networks or towns and retail establishments. Bunge's book *Theoretical Geography*, published in 1962, is one of the most powerful expositions of this approach to geography, whilst Haggett's book *Locational Analysis in Human Geography*, published in 1965, was also an immensely popular exploration of this approach. Using spatial statistics to identify regularities in the patterning of objects (settlements, firms, political boundaries, transport links), this approach to the analysis of space observed spatial patterns and used models to predict changes in spatial patterns. However, in order to provide explanations for changes in patterns, geographers had to appeal to theories about economic or social processes, for example the economic theories of Christaller, Losch and Isard. This way of thinking about space then presented space as a surface upon which patterns created by non-spatial processes were inscribed. Moreover, since these processes operated through time, in this way of thinking time was implicitly presented as active, with space being seen as inert and passive.

Some of the problems with this version of relative space were identified in a paper by Blaut published a year earlier than Bunge's work appeared, in 1961. In this paper Blaut argued that space and time could not be clearly separated: he suggested that space is a relation between events or an aspect of events, and thus bound to time and process. The implications of Blaut's analysis for human geography are that social and economic processes are inevitably spatial and that spatial relationships are always a part of ongoing social and economic processes. Because of the dominance of the spatial analysis approach during the 1960s, the implications of this position did not begin to be explored until the 1970s.

During the 1970s geographers concentrated increasingly on analysing the social and economic processes involved in the creation and perpetuation of inequality. They examined the ways in which groups with different degrees or types of social power used, controlled or operated within space. One aspect of this type of work is that it shows that the significance of a particular location in space derives not from its absolute location and not merely from its location relative to other locations, but from the content of the social and economic processes which link it to or separate it from other locations. Thus, in this type of work, space and relationships in space were increasingly being analysed not only as socially produced but also as an integral part of social processes. One important feature of this change in the view of space is that space is no longer being described as something fixed and absolute but as something that is changed by human activity. Moreover, human understandings and conceptions of space are part of the aspects of space that are being changed by this activity. A further implication of this approach is that there are many spaces which exist simultaneously, all produced by diverse social, political and cultural processes.

During the 1980s the recognition that space is an integral part of socioeconomic organisation became more widely accepted and in 1985 Soja proposed using the term spatiality to refer to socially produced and interpreted space. Whilst there are other meanings of spatiality, this is the one most commonly used at present by human geographers. Its use in current writing implies an acceptance that space is the medium, as well as the outcome, of social action. Another important feature of geographical research during the 1980s and 1990s has been an increasing preoccupation with references to metaphorical space. Increasingly, spatial metaphors are being deployed in much of the academic as well as everyday discussion of contemporary life. Indeed, as Smith and Katz (1993) point out, a common strategy is to use references to absolute space to enhance our understanding of social processes and debates. We refer to social 'location'; we claim, for example, that gender was excluded from the theoretical 'space' of geography in the 1960s and 1970s; we talk of 'mapping' out an 'area' of debate, whilst frequent reference is made to borderlands, boundaries, travelling, 'home' and 'away', and so on and so forth, as metaphors for understanding contemporary life. As Smith and Katz argue, however, this use of spatial metaphor, although often exciting and thought-provoking, also has some significant risks. They point out that we use metaphors to gain understanding by drawing an analogy between something we think we know and understand (in this case space) and something we feel we do not fully understand. Thus the use of spatial metaphors carries the risk that we imply that space is known and fixed – is absolute. However, and as Smith and Katz go on to argue, one of the important implications of the recognition that space is an integral part of social life is that space is always open to contestation by different individuals or groups, many of whom are trying to question and redefine the meanings and boundaries of particular spaces.

Place

Along with space, place is one of Geography's core concepts, and arguably one of Geography's longest traditions has been a concern with people and places, more specifically with what makes places either different from one another or the same. Yet geographers' conceptualisations of place and of the significance of place in geographical study have not remained static.

Underpinning much of the understanding of place in the earliest writings in modern geography is the concept of a region as a geographical area. Based on the empirical documentation of environmental, economic and social phenomena, this geography had the purpose of identifying distinctive and unique areas (regions) of the earth's surface. This tradition is usually traced back to the work of Carl Ritter, whose scientific project was to identify areas or regions with their own particular combination of characteristics and, therefore, their own unique identity. The idea of the uniqueness of place also owes much to the conceptualisation of geography developed by Vidal de la Blache. Working on France, de la Blache conceptualised place in terms of historical strata which combined in unique ways to form local regional identities or personalities.

These ways of thinking about place are ones which are frequently identified as bounded and rooted in particular physical locations and as characterised by fixed sets of social characteristics. More recent ways of conceptualising place, however, start from the premise of increasing globalisation, from the seeming increasing 'sameness' of places, and ask the question 'how is it most appropriate to think about place in these global times?' In these debates the writings of Doreen Massey have been central. Her arguments were initially developed in her thesis on spatial divisions of labour (Massey, 1994), where place was seen as a fluid, historically specific and socially constructed process. Since then, however, Massey has gone on to argue for seeing place in terms of 'a progressive sense of place' (Massey, 1994). Place here is interpreted as the intersection of sets of social relations which are stretched out over particular spaces. Rather than being defined in terms of a particular unique and distinctive location, then, the distinctiveness of place is seen to rest in the combination of social relations juxtaposed together in place and the connections they make to elsewhere. So, for example, Massey thinks of Kilburn not as a bounded area physically located in part of London, but in terms of the sets of social relations which the people who live there are enmeshed in, relations which in many cases stretch well beyond Kilburn itself but which are brought together in Kilburn and constitute its distinctiveness.

In the course of her discussions, Massey (1994) remarks that although many geographers now conceptualise place as an intersection of a number of social relations, each with their own geographies, other definitions of place are used by other groups of people. Her remark certainly encompasses other geographers! In terms of thinking about place, another very important tradition in Geography is humanistic geography. Humanistic geography developed in the 1970s as a reaction against other traditions in the discipline which were working with either absolute or relative space. Humanistic geographers argued that none of this work on space captured the human experience of living in those spaces, and they insisted that this was a crucial absence. In order to rectify this problem, humanistic geographers turned to the notion of place. They celebrated places as spaces given meanings by human feelings. They explored the significance given to places like 'home', 'neighbourhood', 'community' and 'city' by the emotions, memories and habits of individuals. Place for them was a geography made by human sensitivity and creativity; indeed, humanistic geographers such as Tuan (1974) argued that a sense of place was part of what it was to be human.

Humanistic geographers were interested then in place as a part of human identity. However, they tended to assume that there were certain aspects of place about which everyone would feel similarly. The home was one of these: humanistic geographers very often argued that all people search for a dwelling-place in which they can feel at home. While humanistic work in Geography continues, more recently geographers interested in the meaning of places have begun to move away from the undifferentiated sense of human identity used in much of humanistic geography. When humanistic geographers claim that the home is a universally valued place, for example, they did not consider the feelings of those who may have had very bad experiences of 'home': since many homes are places of violence and abuse, not everyone feels that they are places of nurturance and peace. Interested more in social difference and social power relations, many contemporary geographers have begun to consider the ways in which many of the meanings through which places are made are bound into social identities and struggles. A sense of place is often shared by many people; there are, to use Peter Jackson's (1989) phrase, cultural 'maps of meaning' used collectively to make sense of places, and some interpretations of particular places are more influential than others. Cultural geographers argue now that dominant senses of place reflect, in both their form and their content, the meanings given to places by the powerful, for example tourist boards, development agencies, residents' associations and community leaders. Our images of place often depend on theirs. For instance, if you don't live in Scotland (but perhaps even if you do), your image of Scotland as a place is much more likely to echo the hills and glens, bagpipes and tartan, deer and distilleries promoted by the Scottish Tourist Board than it is to reflect the local senses of place of people living in the interwar and postwar housing estates of Glasgow, Edinburgh, Aberdeen or Dundee. However, a consequence of the way in which very specific senses of place are constructed through the particular images and values attached to them by the socially and culturally powerful, is that senses of place are often highly controversial. Other groups may challenge the senses of place produced by the powerful, and cultural geographers therefore argue that senses of place are often also sites of contestation.

Finally, some geographers focus even more closely on how subjective identities are constructed in order to understand how senses of place are made. They argue that identity is made in large part by a contrast to what one is not; we only understand who we are by comparing ourselves to something we claim we are not. That something against which we compare our self is thus something that we ourselves construct, and cultural geographers describe it as the Other. The Other is not understood in its own terms; rather, it is given all the qualities understood as the opposite of those defining it. Some places, therefore, are constructed as otherworldly utopias by world-weary dreamers. But places and their inhabitants can also be given negative qualities by those who consider themselves superior, and contemporary cultural geographers have recently started to examine the ways in which nineteenth-century European geographers exploring colonies did just this as a way of justifying their own presence and power in those places: defining, say, Africa as less civilised and Europe as more civilised legitimated efforts to build European empires in Africa. This process of Othering suggests again that the meaning of places is never absolute. Senses of place are unstable and often contested.

Environment

The environment is the third of the three key geographical concepts with which feminist geographers have engaged. Like space and place, it is also a complex term within geographical debate. As geographers we are initially trained to conceptualise the environment as a physical form which surrounds us. This form can be human-made; the term 'environment' can be used to refer to buildings and streets, as in the phrase 'the built environment'. In more popular usage, though, the 'environment' more often refers to the natural environment. To 'be concerned for the environment', for example, is usually understood as a concern for 'natural' environmental systems under stress, and this meaning is often reinforced by media reports which, as well as expressing general concern for issues such as global warming and deforestation, frequently use shocking images of, for instance, dying forests in Scandinavia, felled rainforests in Amazonia, chemical spillages in lakes and rivers, and so on and so forth. Indeed, such is the media commitment to the coverage of catastrophic environmental incidents such as the nuclear accident at Chernobyl, the oil spilled from the *Exxon Valdez* in Prince Edward Sound, and – as we write – the *Sea Empress* disaster off the Welsh coast, that 'the environment' is firmly on the agenda as a topic of public and political concern. Indeed, the importance of 'protecting the environment' is now widely recognised, in theory if not in practice, by many national governments and by non-government organisations, notably Greenpeace and Friends of the Earth, as well as by smaller-scale organisations such as the anti-roads lobby, currently campaigning against bypass construction in Britain.

As a discipline, Geography has a long-standing concern with understanding the physical processes which structure the natural environment, and geographers have contributed to these contemporary debates about the environment. Some have focused entirely on the environment as a highly complex and sensitive system and tried to understand and predict its response to different kinds of damage. Other geographers, however, have begun to think about the environment in rather different ways. Instead of seeing the environment as something 'out there', with its own logic separate from that of the social, political, economic and cultural systems which may be threatening it, some geographers are now arguing that the so-called natural environment is in fact not separate from the human at all. Not only have humans for millennia impacted on environments in all sorts of ways, but humankind has always, and continues now, to rely on the environment for survival. Consequently, many geographers are now trying to think about the human and the natural as deeply interconnected; indeed, some argue that they are so interconnected that we should no longer think of them as two separate systems interacting on each other at all, but as one.

ACTIVITY

Once you have read through the discussions of space, place and environment, make sure that you can summarise succinctly and for yourself, the various ways in which human geographers have thought about and currently think about these terms.

These arguments about space, place and the environment suggest that they are all complex terms which feminist geographers might engage with in different ways. And indeed they have. As we have already remarked, there is no one feminist geography. Instead, feminists have conceptualised and reconceptualised space, place and the environment in a number of different ways, and for a number of different reasons. Other chapters in this book explore this diversity and difference in further detail.

1.3   The structure of the book and how to use it

Although most of the chapters which follow can be read independently and in no particular order, the chapters also build on and connect with one another. Chapter Two is concerned with examining the various ways in which feminist geographers have constructed histories of feminist geography. Following this, Chapter Three takes one of the central analytical categories within feminist geographers' analyses, 'gender', and explores the various ways in which feminist geographers have worked with this concept. Chapter Four is concerned with debates over method and methodology: it examines the methods and methodologies feminist geographers use to inform their work. Chapters Five and Six then take three of the key geographical concepts – space, place and environment – and explore the multiple ways in which feminist geographers have worked with, challenged and contested these concepts. Finally, in the conclusion to the book we consider some of the broader issues, questions and problems posed by these chapters.

Throughout the book you will find a considerable range of activities for your consideration, many of which come from courses which we currently teach, and a number of which have been designed for this book. These activities can be worked through on an individual basis. Alternatively, they can act as the basis for group discussion, either within a class context or in a self-help group or tutorial. And some could certainly be used in conjunction with the suggested further reading as the basis for project work. As you will find, these activities are located within the text, and they have been designed to enhance and develop your understanding of the key points being made there. As such, we hope that you will not simply pass by these sections of the text, and that you will use them as we have intended. In addition to these activities, a number of chapters also include various case study examples which have been designed to provide you with lengthier illustrations of particular points. Furthermore, the text is also characterised by the use of boxes to highlight key terms, definitions and asides. A final point to note is that the text is characterised by two types of voices, one collective (the voice of particular writing groups) and one individual (where individuals represent themselves and their own views and/or provide reflections and commentaries). In writing this book we have continually had to negotiate the tensions between individuals and their different views, and to accommodate these to produce a collective voice. Admitting two types of voices into our text is the only way in which we felt we could acknowledge these distinctions, as well as reveal their importance in the construction of feminist geographies.

CHAPTER TWO

Contested and negotiated histories of feminist geography

NICKY GREGSON AND GILLIAN ROSE

2.1 Introduction

In this chapter we explore some of the diverse and different ways in which the history of feminist geography has been and may be written. Writing the history of Geography has long been a mini-industry, at least judged by the number of books and journal articles devoted to this subject. Indeed, library shelves are festooned with such contributions, and it is highly likely that you too will have to take a course which discusses some aspects of the history of geographic thought as a compulsory component of your degree programme. One of the features of such courses and books is that precious little attention, if any, is paid to feminist geography: one lecture and an occasional aside are the most that can seemingly be expected. However, as this chapter demonstrates, there is an expanding literature on the history of feminist geography, much of which resonates strongly with contemporary debates on writing the history of Geography (see Box 2.1). The chapter is divided into three main sections, each of which corresponds to a different way of writing history. In Section 2.2 we examine a tradition which we label 'chronicling the development of feminist geography'. This is followed in Section 2.3 by consideration of a tradition which attempts to recover women in Geography. Together these two traditions constitute the main ways to date in which feminist geographers have chosen to represent the history of feminist geography. In Section 2.4, however, we introduce a third option, one which is based on the use of personal, autobiographical testimonies, and which recognises the importance of contemporary debates about multiple voices and situated knowledges. We conclude the chapter by making a few points about the relationship between these three versions of feminist geography's history, and by linking them to debates over writing the history of Geography.

2.2 Chronicling the development of feminist geography

Chronological narratives of the development of feminist geography constitute the predominant mode of representing the history of feminist geography, and

Box 2.1   Writing the history of Geography as the production of situated knowledges

In a recent comment, Felix Driver identifies 'the current enthusiasm for the writing of new histories of geography (as)...one of the most striking developments across the discipline' (Driver, 1995: 403). Much of this enthusiasm is indicative of the debate over one of the most recent books to appear in this field, David Livingstone's *The Geographical Tradition* (1992). In this Livingstone challenges many of the features associated with the vast array of textbook histories of geography, specifically their presentism (telling history in terms of enabling an understanding of the present), their construction of history in terms of great names, and their internalism. Geography's story, he argues, '(is usually) written by geographers, about other geographers, for still other geographers' (1992: 4). And these stories rarely feature discussion of 'social context, metaphysical assumptions, professional aspirations or ideological allegiances' (1992: 2). In contrast, Livingstone argues for a contextualist history of geography, one which acknowledges that the nature of geography has always been contested and negotiated; one which recognises that geography has meant different things for different people in different places; and one which focuses on accounting for how and why particular practices of geography get to be legitimated at different times and in different places. The phrase commonly used now to refer to this range of issues is 'situated knowledge'. Knowledge is never pure but is always situated in the complex and sometimes contradictory social locations of its producers and audiences. In a similar vein, Driver continues the emphasis on 'situated messiness', arguing that efforts to portray the geographical tradition need to focus on the heterogeneity of geographical knowledges, as well as 'the material circumstances through which ideas take root'. Here he calls for a construction of unfamiliar histories of geography, as opposed to the familiar ones of standard surveys; histories which recover lost and/or marginal figures, and which construct alternative, counter-traditions. Using the examples of Stanley, Yorke and Leade in 1872 to illustrate the diverse, fractured and contested nature of geographical knowledges at any one time, Driver argues: 'White, male and British they all were...but this is hardly sufficient (basis on which) to conclude that their geographies were either as singular or as confident as we, in our singularly confident moments, are tempted to presume' (Driver, 1995: 411).

References

Driver, F. 1995. Sub-merged identities: familiar and unfamiliar histories. *Transactions of the Institute of British Geographers*, 20(4), 410–413.
Livingstone, D. 1992. *The Geographical Tradition*. Oxford: Basil Blackwell.

are to be found in a wide variety of journal articles and texts (see the Bibliography). Typically such narratives focus on the changing conceptual orientation of feminist geography; they outline the key ideas and concepts which feminist geographers have worked with over time. Furthermore, at least until recently, they have tended to portray these changes as sequential, progressive and universally true. Indeed, we could summarise this stance as presenting a story line which tells how feminist geographers first did this and then did that.

One of the best examples of this style of writing feminist geography's history is to be found in Sophie Bowlby's 1992 article 'Feminist geography and the changing curriculum'. Written for the journal *Geography*, and therefore aimed at sixth-formers and their teachers, this article presents the history of feminist geography in terms of a threefold progression, itself demonstrative of intellectual growth and maturity. Starting with feminist geography as the geography of women, Bowlby argues that work then moved on to consider the processes by which 'men (maintain) more power in society than women' (p. 349) and, more recently, has changed again as the importance of differences between people and cultures is increasingly recognised within feminist geographers' work. A similar story, albeit told with more twists and nuances, is provided by Sophie Bowlby, Jane Lewis, Linda McDowell and Jo Foord in the text *New Models in Geography* (1989). Here we find a brief outline of the history of feminist geography, one which emphasises the main strands highlighted by Bowlby. The narrative is one which charts a progression from work on the geography of women, through studies which emphasised explaining patterns of gender inequality, to work on gender identities. This same narrative of progress is used to shape summaries of feminist work in urban/social geography and industrial/regional geography, and the article concludes with some brief comments on future directions.

ACTIVITY

Read both the above articles and identify the differences and similarities between their stories.

Bowlby, S. 1992. Feminist geography and the changing curriculum. *Geography*, 77, 349–360.

Bowlby, S. *et al.* 1989. The geography of gender. In Peet, R. and Thrift, N. (eds), *New Models in Geography*, Volume Two. London: Unwin Hyman, pp. 157–175.

Articles such as the above, together with others of the same ilk, are important, not just for what they might or might not tell us about the development of feminist geography, but for what they reveal about the diverse ways in which we may choose to present this history. Both these articles, for example, are presentist: they tell the history of feminist geography in terms of understanding its present. They focus on those aspects of past work which led directly to current emphases. Ideas and pieces of work which either don't quite fit in with this chronological story or led elsewhere are omitted from the grand narrative. In addition, these two articles are imbued with the notion of history as progress; what comes after is considered to be an advance on what went before. In accounts such as these, feminist geographers are seen to be moving forward, and just as importantly, doing so fast, in itself something which can be seen to be both supportive of claims of vibrancy and dynamism and illustrative of the intellectual maturity of feminist geography. Finally, we can note that both these articles reveal how the version of history 'spun' varies according to the supposed audience: a schools focus produces a simplified, accessible chronology demonstrative of intellectual growth but with plenty of examples; that for undergraduates is more 'fleshed out' and complex. The writing of feminist geography's history, therefore, is shown by these articles to be an inherently strategic endeavour, one in which different geographical audiences are persuaded in different ways of the intellectual credentials of feminist geography.

At the same time, however, we need to make some connections between this mode of writing the history of feminist geography and some of the points made in Box 2.1 on writing the history of geography. There we saw how critiques of much of the history of geography literature draw a distinction between 'internal' and 'external' or 'contextual' histories, and problematise the homogenising tendencies of grand narrative. Such criticism can (and should) be applied to representations of feminist geography's history, notwithstanding the location of these histories in opposition to histories of geography. Accounts such as those which chronicle the development of feminist geography, then, are internal narratives, and need to be recognised as such. Their concern is with constructing a lineage of ideas, albeit a lineage counterposed to that of mainstream geography. And undeniably this lineage fails to do justice to the great variation in ideas which have been around in feminist geography at any one period or place.

To an extent such criticisms have been taken on board in some of the most recent retrospectives on feminist geography. Linda McDowell, for example, in a two-part review article in the journal *Progress in Human Geography* (1993a, 1993b) identifies three traditions within feminist geography which she sees as existing to a degree simultaneously, and which she tries to link to broader developments in feminist scholarship and in geography. Despite acknowledging heterogeneity, though, the focus in these papers is still predominantly 'internal' and there is a sense too (which McDowell recognises) in which the simultaneous traditions slide through the narrative into chronological advancements. In Chapter Three in this book we too attempt to go down the route of writing the development of ideas within feminist geography, but in a way which takes heterogeneity seriously. Quite how successful we have been in this endeavour we leave you to decide.

SUMMARY

- Feminist geography's history can be told in terms of the sequential progress of feminist geographers' ideas.

- Such versions of history are presentist and internal, and portray a homogeneous vision of feminist geography's development.

2.3   Recovering women in geography

A second strand in writing the history of feminist geography is one which attempts to recover women in Geography, and it is to this that we now turn.

In contrast to the approach adopted by the tradition of chronicling the development of feminist geography, those who choose to represent the history of feminist geography in terms of recovering women in Geography contest both the traditions which are presented as Geography and the lineages on which these traditions are based. In this work the construction of geographical traditions is conceptualised in terms of a constant process of inclusion and exclusion. As Gillian Rose, writing in relation to the writing of traditions, states: 'Certain people or kinds of people are included as relevant to the tradition under construction and others are deemed irrelevant' (Rose, 1995b: 414). In terms of geographical traditions there is little doubt that these are characterised by what Rose describes as 'the persistent erasure of women'; women are written out of Geography's histories as Geography is constructed in terms of paternal lines of descent. For much of the time, then, geographical traditions are constructed as internal histories, by 'Would be great men cit(ing) men already established as great in order to assert their maturity' and by 'rebellious sons construct(ing) paternal lines to revolt against' (Rose, 1995b: 414). However, even in external, contextual histories of the type produced by David Livingstone in *The Geographical Tradition* (1992), women are invisible. Indeed, in this book, widely cited as the best in its field, only two women receive a mention in half a millennium of Western geographies!

ACTIVITY

As well as being written about in the pages of academic books and journals, Geography's history is frequently represented visually in the buildings within which Geography is taught. Figures 2.1 and 2.2 comprise montages of portraits, busts, photographs and other assorted artworks which are displayed prominently on the walls of various UK Geography departments. You may be able to identify your own department within these images. What do these montages suggest to you about how the history of geography is constructed and represented? And how does the visual representation relate to the written? In answering these questions you might like to think about who is included in these images and who is excluded and what type of traditions are portrayed (there is evidence here for paternal lines of descent, for paternal 'families' and 'kingmakers'). Does your own department go in for the same type of visual display? If it does, have you ever given these images much thought? If it doesn't, why do you think this might be?

For a growing number of feminist geographers, an appropriate response to this persistent erasure of women from geographical traditions has been to reveal this as such and to show that women have been involved in the construction of geographical knowledges. One of the best examples of this is to be found in recent work which has looked at Victorian women travel writers. A glimmer of what is contained in their writings is to be found in an article by Mona Domosh (1991a), whilst more detailed studies have been produced by Mary

Figure 2.1   Representing Geography's history I

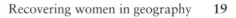

R.N. RUDMOSE BROWN

D.L LINTON

SIGN FOR BUILDING NAMED AFTER
PROF H. DAYSH

PROF H. DAYSH

PROF H. DAYSH

Figure 2.2   Representing Geography's history II

Louise Pratt (1992) – not a geographer – and Alison Blunt (1994). As interesting as the 'recoverings', however, has been the response to them. Before making her case for the contribution of women to the construction of geographical knowledge, Mona Domosh (1991a) criticises David Stoddart's book *On Geography* (1986), using many of the arguments outlined above. Fairly unexceptional, one might think. His response to this criticism indicates otherwise, and is highly illuminating. Indeed, he states quite baldly, 'the simple fact is that none of the persons Domosh discusses has anything to do with my themes: there is therefore no need to mention them' (Stoddart, 1991: 484). Not only are women travellers dismissed here because they don't fit in with the tradition which Stoddart is constructing (modern scientific research, for which competence in fieldwork and exploration, defined in accordance with scientific principles, is deemed essential) but, as Rose says, 'the process of excluding women from this geographical tradition vanishes too' (1995b: 414).

ACTIVITY

Read Mona Domosh's article, Toward a feminist historiography of geography, *Transactions of the Institute of British Geographers*, 16, 95–104. Then, read the extracts contained in Reading A following Section 2.5. These are taken from Stoddart's reply to the initial article and Domosh's response. What are Stoddart's objections to Domosh's arguments? How would you respond to the points which he makes? Do you think any of these arguments are accurate criticisms of Domosh's paper? What about any of the others? And what do you make of the points which Domosh makes in reply? What do you think about Stoddart's apparent understanding of the term feminist historiography?'

As well as exposing the way in which the disciplinary agenda(s) in geography is or are gender specific, the recovering women tradition of writing history reveals the multiple ways in which the tradition of feminist geography can be constructed by feminist geographers. In some respects there are similarities with the approach discussed previously. But there are also significant differences. Here, as there, there is an attempt to construct lineages. But this time it is an embodied lineage, rather than a lineage of ideas and concepts; it is a lineage which is attempting to uncover geographical 'foremothers'. And there is a difference too in that presentism is challenged (at least partially). Here history is being written not in terms of understanding contemporary circumstances, but in terms of forgotten or erased knowledges, of lines of writing and analysis which were marginalised, overlooked or bypassed by other (masculine) lines of enquiry. Such strategies of writing history then provide ample testimony to the heterogeneity (as opposed to homogeneity) of geographical knowledges and to the legitimacy of non-hegemonic geographical knowledges today.

But there are criticisms which can be made of this approach to writing history. One of the features of this approach is that it is celebratory of women and their difference from male geographers. There is a sense in which women like Victorian women travellers are being celebrated not just for what they did, but because they are women, and because they are women who were writing things which we, from our position today, can label as geographical knowledges. Such a stance comes perilously close to being both essentialist (see Box 2.2) and presentist. Furthermore, few of these celebrations pay much attention to the fact that most of these women travellers were white, middle class, able bodied and of independent financial means. Neither do they reflect much on the centrality of this positionality both to enabling the production of these geographical writings in the first place and in the very nature of these writings.

Box 2.2   Essentialism

Following Diana Fuss (1989), essentialism is most commonly understood as a belief in the essence of things, as a belief in the existence of 'invariable and fixed properties which define...a given entity' (p. xi). In feminism such arguments appear in accounts which appeal to notions of a pure femininity and in work which argues for the existence of a female essence beyond the boundaries of the social. Such accounts frequently assume that sexual difference is prior to social differences, with the latter being mapped onto the former. Essentialism also figures prominently in writing which argues in terms of universal female oppression, and in work which explores the potential of a feminine language. In contrast to essentialist arguments are those writers who take an anti-essentialist or constructionist position. Such writers criticise the essentialist position by showing, for example, the difference between what is understood by the categories 'man' and 'woman' in different societies. They point to the importance of historical and cultural context, arguing that what we understand by essence is itself a social construction. For anti-essentialists, it is the social construction of ideas, concepts and categories of thinking – through systems of representation, material and social practices, and discourse – which matters.

Bibliography

Fuss, D. 1989. *Essentially Speaking: Feminism, Nature and Difference.* New York: Routledge.
Graham, J. 1990. Theory and essentialism in Marxist geography. *Antipode*, 22(1), 53–67.

ACTIVITY

Readings B and C give a flavour of some of the concerns of one of these women travellers, Mary Kingsley, and offer two contrasting ways in which these writings might be interpreted. As you read these think about whether these arguments are essentialist or anti-essentialist, and whether they celebrate Mary Kingsley. Then try to work out the main differences between Mary Louise Pratt's and Alison Blunt's readings of Mary Kingsley's writings. How might these differences relate to different ways of thinking about space? In answering this question you might like to think about how the two authors seem to conceptualise space (think here about literal space, metaphorical space, spatiality, spaces of complicity/resistance and the ways in which the subject is represented as constructing and contesting space).

ACTIVITY

The criticisms outlined above, whilst hardly commonplace, are ones which have at least begun to appear in the literature. In the course of writing this chapter, however, one of us (provocatively) posed the question: to what extent might the recovering women tradition be open to interpretation as constitutive of an oppositional tradition in which would-be great women comment on, and recover, their great female precursors? To what extent do you agree or disagree with this idea? In thinking through your response you might like to bear in mind the following points: what is the difference between recovering women and men in a male-dominated discipline and society? In your opinion, would any aspiring great woman choose nineteenth-century women geographers as a way to conquer the discipline? Might this subject area enable the construction of maternal lines which rebellious daughters could use to disrupt the certainties of paternal lines of descent?

SUMMARY

- Feminist geography's history can be told by recovering Geography's 'foremothers', those women whose writings have been erased from masculine constructs of the geographical tradition.

- Such versions of history celebrate women, and women whose writing can be labelled as 'geographical'. Such accounts can be presentist, are very often essentialist, and are also internal histories.

The contemporaneous existence of the above two versions of writing the history of feminist geography testify to the complexity of writing the history of feminist geography and to its contested nature. Writing history is never simple, neither is it straightforward. Instead it is messy and imbued with our visions of the present. Indeed, we construct histories both to construct traditions and to legitimate ourselves, and because we differ we produce different histories, for different reasons.

The above strategies, however, are not the only ways in which it is possible to write a history of feminist geography. Increasingly, in the context of debates about the making of knowledge in Geography and elsewhere, feminist geographers have become concerned about the homogenising narratives which structure and produce both the above traditions and the extent to which these fail to provide a fully contextual interpretation of the history of feminist geography. Moreover, although implicated in feminist geography, most of those writers producing these histories neglect to mention the nature of their own involvement in feminist geography and the ways in which they may or may not be implicated in such projects (note that in Chapter Three, where we attempt our own version of Section 2.2, we make specific reference to this). In the next section, therefore, we disrupt these tendencies by introducing several 'views from somewhere', personal testimonies which admit multiple voices and different stories, and which reflect on the institutional context(s) and social relations which constitute situated histories. We begin the section, however, by locating this version of writing history within contemporary debates on writing.

2.4   Writing personally: autobiographical histories of feminist geography

Writing personally is not something which comes easily to academics, at least in an academic setting, and to preface this section we felt it would be useful to consider why this is so. Much academic writing, certainly the writing which you will have encountered in many of your courses so far, and which many of you will have been encouraged to emulate in your essays and course papers, is characterised by a dispassionate, distant, disembodied narrative voice, one which is devoid of emotion and dislocated from the personal. In contrast to this, writing which is personal, emotional, angry or explicitly embodied is implicitly (and often explicitly) portrayed as its antithesis: something which (maybe) has a place in the world of fiction and/or creative writing, but which, quite definitely, is out of place in the academic world. The pre-eminence of the dispassionate, distant and disembodied voice is something which has been challenged repeatedly in human geography over the past 20 years or so, for example by those who initiated the first critiques of positivism. Humanistic geographers, for example, argued that such a writing voice simply could not evoke the creative and emotional senses of place which they argued were fundamental to 'place' as a geographical concept. Marxist geographers suggested that the supposed neutrality of that voice served merely to hide its lack of critique of capitalism; it was a voice whose apparent objectivity masked a conservative and reactionary set of values. More recently, feminist geographers have also criticised that voice as expressive of certain kinds of masculinities (see, for example, G. Rose, 1993). Particularly in the academy, where cool and calm rationality is the desired norm of both behaviour and debate (which is not to suggest that this is what actually happens!), to be masculine often means not to be emotional or passionate, not to be explicit about your values, your background, your own felt experiences. Feminist academics wishing to challenge those exclusions from the written voice of Geography find themselves in a dilemma, however. For if academic masculinity is dispassionately rational and neutral, writing which is overtly emotional or explicitly coming from a particular personalised position is often dismissed as irrational, as too emotional, as too personal – as too feminine, in other words. Thus feminists who want to assert the importance of the emotional in their work, or feminists who want to acknowledge the personal particularities of their analysis, run the risk of being read as incapable of rational writing, of merely being emotional women whose work cannot be universally relevant. The response of the academy is all too often to trivialise, marginalise and even ridicule those who do not write in its conventional style, a response which acts to silence some and to determine who gets heard.

An excellent illustration of this tendency is to be found in a recent issue of the journal *The Canadian Geographer* (1994), which contains the Wiley Lecture given by Peter Gould to the Canadian Association of Geographers in 1993, three responses to this, and Peter Gould's reply. Angered by Peter Gould's remarks regarding, amongst other things, feminist geography, Linda Peake writes an angry and personal response to this, which is subsequently trivialised, and marginalised, in Peter Gould's reply. Reading D gives some flavour of the proceedings.

ACTIVITY

Read the extracts contained in Reading D or, if you have the time, the original lecture, the three responses and the reply. How do you respond to Gould's remarks regarding the feminist challenge, to Peake's response and to Gould's final remarks? Where would you position yourself with regard to these two protagonists? To what extent are contrasting ways of writing central to understanding the differences between Peake and Gould, and how do these extracts illustrate the points we made above concerning the relative acceptability of dispassionate (masculine) writing, as opposed to personal and angry (feminine) writing, within the academy?

Writing personally, then, is a high-risk strategy for academics. Small wonder that many of us found writing a personal testimony exceedingly difficult, and why some contributors include academic references in their testimonies. In strategic terms, however, writing personally has to be reclaimed: if we are to challenge the dispassionate, disembodied, distant, masculine voice and to contest the ways in which this marginalises and trivialises the personal as feminine, then we have to both expose the former and reclaim the latter, for writing personally is central to notions of situated knowledge and to fully contextual versions of writing history.

In what follows in this section, then, in line with this strategic endeavour and with the explicit intention of illustrating the importance of diversity and context in the history of feminist geography, we present verbatim, and in no particular order other than alphabetically, some of the testimonies which we received. A further testimony, written by Janet Townsend, also appears in Chapter Three (Section 3.3) which you should read in conjunction with those that follow here. Potential contributors were contacted via both a general call in the WGSG Newsletter and through a mail shot of all members of the study group. Both requests asked people to write a short 'history of the present', a piece about themselves and their engagement with feminist geography, and we received responses from both those involved in writing this book and those who were not.

Claire Dwyer

When faced with the request to write a personal testimony as a feminist geographer my first reaction was to wonder: Am I a feminist geographer? Would I be accepted as a feminist geographer? Henrietta Moore (1994: 8) writes about the same anxieties of being exposed as 'not a proper feminist'! So I have tried to think about when I was first enthused by feminist ideas, and how I came through various routes to a sense of feminist consciousness.

As an undergraduate I had little exposure to feminist geography – beyond one lecture as part of a social geography course. My interest in feminism came through my activities outside the geography department where I was involved in the student action group Third World First. It was here that I became interested in questions of women and development which provided an impetus for my undergraduate dissertation in India. My dissertation evaluated the success of a rural health care programme by interviewing women about their water use practices. The research was developed within a feminist framework (although I would not have defined it as such then) which focused on the role of women as knowing subjects who made rational decisions within a limited framework of opportunities.

After university I completed a Postgraduate Certificate of Education where I was introduced to theories about gender and education and became concerned to develop strategies for learning which might be defined as feminist. Putting these ideas into practice in the next few years' teaching was not always easy! Certainly it was as I worked as a secondary school teacher that I first became really aware of the implications of gender inequalities. Yet it was also through my experiences within the gender working group at school that I realised some of the potential for feminist teaching strategies for change.

In 1990 I travelled to the USA to begin a Master's Degree in Geography. Although there was little feminist geography I was free to attend classes in any department and so found myself in my first semester in an education class on 'Gender, Culture and Education'. This course was my first introduction to academic feminism but was also an intoxicating conversion into a feminist consciousness which gave me a language to explain things I had previously been unable to name. Such was my evangelical fervour that in my first summer back in the UK doing research I organised feminist reading meetings for a group of friends – eager to share my new found knowledge!

The first course was important not only because it provided me with an opportunity to read things I had never read before but also because of the 'consciousness raising' which I shared with a diverse group of women most of whom had, like myself, little grounding in feminist theory. The next semester I was to take a course in the English department on 'Gender and the Culture of Television' which was to extend and provoke my feminist consciousness as the class struggled around questions of difference and diversity through the challenges of a number of African–American students to a predominantly white audience.

These two experiences were, I think, fundamental in creating for me a sense of identity as a feminist geographer. It was through these classes that I found a theory I cared about enough to do academic work as well as a theory which provided an explanation for my day-to-day struggles as a teacher. Returning to the UK to work on my PhD my support networks as a feminist geographer have been more fragmented and I have had to work more independently to educate myself about feminist theory. However, the Women and Geography Study Group provides, if only periodically, the same source of support and enthusiasm as my weekly feminist theory class. In terms of my work I have been stimulated to research a PhD which centres around the issues of feminism and religion as both an academic exercise but also as a personal journey of making sense of two important influences in my own life.

Nicky Gregson

For me, the decision to write this testimony has been a long and difficult one. Writing in this way is demanding, difficult and (potentially) damaging, both for those of us who choose to write thus and for feminist geography more broadly. I've therefore chosen to preface my reflections by making explicit some of my own dilemmas in writing this piece. I have, for example, worried intensely about just what to include here and what to exclude. I've worried about how best to write what I'd like to say. And I've also worried about how this piece might be 'read', about how whatever is said might be interpreted by others, specifically by other feminist geographers and by individuals outside of feminist geography. In the face of all this angst I've frequently wondered whether certain things might just be better left unsaid. Moreover, I've also questioned whether in writing ourselves we might be laying ourselves bare (metaphorically speaking), particularly in relation to the departments (and persons) within which (and with whom) we work on a day-to-day basis. As if writing about something in which I am implicated (albeit in a small way) and represented, and which I have had multiple feelings towards (including ambivalence, marginality, alienation, warmth and strength) isn't hard enough! In the end, however, my belief in the project of writing ourselves has overcome the dilemmas. The product is an account which has many deliberate silences and absences, yet which I hope has neither excluded so much as to be anodyne, nor included material which might seem trite and/or overly idiosyncratic. It is also an account which disrupts the conventions of linear narrative. When I think about my relationship to feminist geography, it is impossible for me to write this as a straightforward chronological account. Instead, this relationship is fluid, neither one thing nor another, something which reflects personal influences, institutional contexts, identity politics, my interpretations of 'being an academic'...And all of them simultaneously. I've therefore chosen to write this testimony as three 'takes' on feminist geography, each of which reflects differently on what feminist geography has meant for me.

First encounters/daily lives

The first time I heard about anything called feminist geography, as opposed to feminism, was in 1983, when Janet Townsend gave a seminar in Durham. I was then a postgraduate student working on my PhD thesis, on the emergence of capitalist class relations in Northwest England. On reflection, this was an agenda which reflected several things: the strength of political economy within Durham at the time, or more accurately, the critical influences of a few key individuals; my own intense interests in class relations and the contexts which shape these; my love of sixteenth/seventeenth century social/economic history; and a passionate desire to prove (to myself) that it was possible to produce an historical geography which wasn't about the morphology of rural settlement! The representations of Cumbria which I was dealing with in this project were largely the creation of Lord William Howard, his male descendants and their various (male) estate agents. As for women, well, they played a few 'walk on' parts (largely as widows), but sheep and beef cattle were far more important! I'm ashamed to say that I don't recollect precisely what Janet had to say that day, although I do recall spending quite some time after the seminar thinking hard about the place of feminism within the academy. What does stick in my mind, though, from this first encounter with feminist geography is the performance: as Janet entered the room (a room festooned with male tribal relics, Middle Eastern carpets and various other symbols of Geography's imperial past, and peopled with many male staff who did not usually attend departmental seminars), a couple of male staff performed two symbolic rituals. One pulled the chair out for her; the other helped her with her poncho. Looking back, this encapsulated for me the extent of the battle facing feminist geographers. It also filled me with admiration for Janet's courage and determination, both to speak out in this context and to work in it. Now, some twelve years on and as I enter my own department on a daily basis and face similar (if less explicit) battles, I continue to draw strength from her example.

Becoming 'a feminist'

In late 1985/86, against a backdrop of swingeing cuts in higher education, few – if any – academic vacancies, and highly limited funding opportunities in historical research, I took the major decision to switch academic fields and move out of historical geography into contemporary human geography. By then I had moved, via Exeter, to Newcastle. Since then feminism has occupied a central place within my academic writing and thinking. This, I think, was inevitable. For almost as long as I can remember, 'women's issues' and 'women's experiences' – if not always an explicitly feminist consciousness – have occupied a significant, defining part of my life and identity, albeit inextricably interwoven with class. Some of my earliest recollections are of both my mum's and nanna's struggles to feed and clothe our households on next to nothing: of the skill and ingenuity which went into food preparation, making and mending clothes, and home decoration; and of the immense amount of time and labour which this absorbed – and not just theirs. Before I went to school I can remember routinely helping out on Monday's wash: we had a highly unreliable twin tub which, if left unattended, emptied gallons of water onto the kitchen floor. One of us children had to hold the outflow hose over the sink, whilst mum and a neighbouring mum dealt with the spinner! And shopping was a major expedition – no car and no bus meant we all (i.e. mum, brother and me) had to walk to the supermarket and then carry the weekly shop home (a round trip of three miles). Later on, once I went to secondary school, the debates shifted away from domestic labour (although I still continued to have to do this) to reproductive and body politics. None of us at that time would have labelled ourselves 'feminist', yet we all discussed at length 'teenage pregnancy' and abortion (usually as this impinged on the lives of individuals in our class), as well as the Pill, and all of us agreed with 'a woman's right to choose'. We therefore took on board many of the demands of second-wave feminism, without then seeing the necessity of the label 'feminist'. Later still, these debates about reproductive politics became even more complex: in my own case they revolved around the medical treatment which my mum received over a four-year period of being labelled 'menopausal'. Indeed, the nightmares of living with anti-depressant therapy, through withdrawal and to an eventual hysterectomy were a turning period for me. In short, they encapsulated in an intensely personal way the inequalities which women experience, as well as the incredible difficulties which women face in contesting these inequalities and in challenging society's representations of Woman. 'Feminist' became a label I was happy to adopt, aptly through reflecting politically on the personal. So I guess feminism in its wider political sense has always been 'there' for me, shaping me and positioning me. Seeing social science as about both understanding the world of which we are part and seeking to change this 'for the better' has meant that feminism from the start has defined and informed my research agendas, and that it continues to do so today. Moreover, that some of the projects which I have been involved in over the intervening years have been (and are) concerned with domestic labour and medical representations of Woman is further testimony to the extent to which 'the personal' continues to shape my interpretations of being an academic.

Relations with feminist geography

My relationship to feminist geography (as opposed to feminism) is one which has fluctuated considerably over the past ten years. An overriding constant in this has been my continued belief in the power and importance of the critiques which feminist geographers bring to bear on geographical knowledge. But at various times I have felt to varying degrees disconnected, ambivalent and even disheartened about this thing called feminist geography. Some of this I think – certainly the disconnections and ambivalence – stemmed from being outside the project that was feminist geography in its earliest days. I wasn't 'in' on *Geography and Gender* and this made a difference. Part of it too, however, was about my long-standing problems with Geography. My work has always been interdisciplinary in nature, and I found it highly constraining (not to mention uninteresting!) to put a spatial gloss on whatever I was interested in in order to legitimate it as 'Geography'. So, whilst the feminism was always central for me, the geography was always more peripheral. I guess therefore that I was never entirely happy with the label 'feminist **geographer**', as opposed to the label '**feminist** (geographer)'! The disheartened phase is one which I can pinpoint more specifically. This was 1989/90, a period which saw feminist geographers' first attempts to debate the challenges of post-modernism and which – as elsewhere – produced painful discord, sharp differences and intense disagreement. At the time I felt that feminist geography appeared unable to cope with this: we seemed to be imprisoned in a land where we either skated round our differences or (occasionally) erupted into acrimonious exchange and confrontation (Penrose et al., 1992). With hindsight I think that it was almost inevitable that such differences would prove difficult for a group such as the WGSG (with its strong tradition of friendship, support and consensus) to work through. I like to think, however, that the 1990s have seen the WGSG learn to work with and through difference. Feminist geography is no longer (probably it never was) a space of intellectual and political consensus. Instead, under the influence of different feminisms and transformations elsewhere in Geography (notably the cultural turn and the sexuality and space network), it has diversified, I think for the better. These changes, in conjunction with the emergence of more creative ways of thinking about 'the spatial' , have meant that I have once more felt myself drawn to feminist geography and the writings of those who are happy to label themselves feminist geographers.

Elizabeth Kenworthy Teather

The geography I write best is the geography I'm involved in – the geography of exile, of being uprooted and rootlessness, of placemaking and homefinding. If it's feminist at all it's instinctive and not informed by much knowledge about feminist theories, because the first opportunity that I've had to read about this area was in 1994, on my first study leave. And by then I was fifty-one!

I didn't even want to do research on women when I was appointed to a post in a Geography department, somewhat unexpectedly, in 1988. It had been very difficult to get a job in this small town of 22,000 people, and I had recently completed a second Honours degree, this time in Drama, and would have been just as happy to get a tutorship in the Drama Department... However, the only job I could find was teaching in a private school, and when a tutorship came up in Geography and Planning I applied in some desperation.

Since completing my PhD at UCL in 1970 I had not had any contact with academic geography, although I had kept up my membership of the GA (Geographical Association) and IBG (Institute of British Geographers) and enjoyed reading the journals and reading about the eminent positions to which some of my former fellow students had risen.

So there I was in 1988, a very junior academic at forty-five, expected to begin some research or other. And, eventually, research on women was simply irresistible, because there was so much to be done! But what really got me started was when Linda McDowell accepted, almost by return of post and with an encouraging letter, a brief article I submitted to *Area* in 1989. I owe a lot to Linda, because since then I've never looked back!

It's been good fun getting out in the bush and talking to rural women, and getting their stories into the geography and social science journals – and eventually doing comparative work in New Zealand and Canada. I have to report that even activist female farmers of the Ontario variety seem to feel most comfortable and relaxed talking over the kitchen table or while preparing a meal. Dare I admit that my recipe collection increased during my 1994 research?

So what feminist position have I arrived at? Others have written about my research from a feminist theory point of view. Currently, I'm finding plenty of scope in other theories, notably structuration. My most encouraging colleague has always been my husband, who isn't a geographer. It's he who is responsible for my having to cope with being uprooted again and again. Next to his comes the encouragement I've received from some of the women I've befriended while researching them. With honourable exceptions, most of my predominantly male colleagues have tended to see me as working in a peripheral and insignificant area, which of course it is, power lying in the hands of men, especially in macho Oz. What I find hardest, professionally, is the urge my colleagues have to find 'new, young blood'. I was fifty when I reached the dizzy ranks of tenured lectureship, and I reckon I'd earned it. New – enthusiastic, hard-working – blood doesn't have to be young, dammit.

Funny, isn't it, that my research subjects are about my age. Here I am, finding out about experiences similar to my own through my research, i.e. looking at how women's contribution to Australian life can all too easily be overlooked. I don't want to be overlooked, nor do I want them to be. Which all seems pretty feminist to me.

Nina Laurie

The legitimate thing

My Geography credentials are impeccable: I get on aeroplanes to do research, I am often seen in the corridors in 'ethnic' garb and I have a mountain named after me in the southern Andes. By virtue of these circumstances I am seldom asked the 'is it geography?' question which many of my colleagues interested in gender and social relations constantly have 'to field' (pun intended). On the rare occasions when the academically sceptical wonder about my 'geography' their fears are soon allayed by my 'I do research on gender and development' reply – it seems the potential for lucrative consultancies can also be a significant redefine of disciplines these days.

My feminist markings are equally as good – I have chipped teeth gained from an early playground battle for refusing to take part in 'kiss chase' and an on-going interest in challenging the gender dynamics which discriminate against people who do not want 'to play' games when they have no say in setting the rules.

With such an introduction it must be said that I also come from a long line of feminist geographers. While at an Erasmus conference in Barcelona in 1995 I was sitting at a table with Janet Townsend, Nicky Gregson, Uma Kothari and Janet Momsen: Janet T taught Nicky and Nicky taught me (as did Janet M), if one of my students had been with us she would have been 'the fourth generation' – our own version of 'to boldly go'. Into such a line come other influences. Both my supervisors were women: Audrey Kobayashi and Ann Varley; and when Janet Momsen left to go to the States Uma Kothari did her job for a year, then when Uma left to go to Manchester I replaced her. It could be said we have a successful 'old girls' network' – I have now become the permanent fixture in 'the woman's job' in my (our) institution.

The motivation thing

My interest in Geography and feminism probably started with my mother, who always told me we (women) could do whatever we wanted. She had been in the Merchant Navy before she married, and travelled extensively, and yet throughout my teenage years I watched her being ground down trying to make ends meet and employing various 'survival mechanisms' in the process. As a single parent she rose early to pick strawberries in sweltering plastic funnels; from there she went directly to work in an illegal clothing factory and in the evenings she waitressed in that male bastion 'the Masonic hall'. So now, when I study gender and development, sweat shops, double/triple burdens, and the feminisation of the labour force, it all seems quite familiar, not the exotic Other at all.

The focus of my research and teaching is often on resistance and I suppose I am inspired by the way in which my mum resisted. Her interest in me as a person and her wacky stubbornness were the hallmarks of her feminism.

The difficult thing

So far, then, my history reads more or less like the 'fairy tale' feminist geographer's path: a linear progression, surrounded by travel and great role models – strong, independent, witty women. However, things are never that simple, are they? Many of the lessons I learned along the way were hard ones, mainly centred around 'growing up' and realising that women on pedestals have a tendency to topple off. It took me a while to realise that 'sisterhood' is not a universal pink woolly rug you pull up around you to keep warm and snuggle into. The painful side of the old feminist slogan 'the personal is political' is that disagreements with colleagues (other feminists) can become a bit like falling out with your friends. Especially when the disagreements are over 'who said what and to whom, when and why' rather than academic debates. Some of the people I have admired most have 'let me down'; their personal lives, interactions with people, research agendas, departmental politics, etc., have not always been consistent with their espoused feminist values. This is hard to handle and at different points it made me want to give up the whole thing and stay somewhere warm and hot where people speak Spanish, drink wine and live out other geographies which from a distance and through rose-coloured glasses, seem easier.

However, it has taken me a while to realise that I was suffering from the pop star syndrome, expecting individuals to live up to the music they make. Sometimes they do, but sometimes they don't. Two things have helped me accept and understand this. The first is I got a paid job in a university. As a result I saw universities and the teaching project in a different (less rosy) light; departments with few women; corridors of power full of men in grey suits (you know the ones from the two rails in Marks and Spencer with the slightly flared trousers... next to the zip-up cardigans with suede patches); and finally the experience of committee meetings. It was a shock. I somehow expected the social sciences to be a hotbed of radicalism full of people who were like my friends. It wasn't like that at all, at least not initially. Linked to these experiences was the surprising realisation that there were dangers associated with teaching. I believe in teaching, I like it. I like interactive classes, questions and challenges. I also believe that the quality of teaching delivered should be good. There are two dangers associated with this belief, however. The first is that if you espouse such a philosophy, your teaching load goes up, you get labelled as a teacher. You therefore cannot 'get on with' research. When you try to do both you head for a breakdown. The second danger is from the students. Sometimes I think that you aren't interested, or you think you know all about 'development' (which is what I teach) because you have watched a programme on Channel Four. You often find open teaching methods 'cringey'. Sometimes you don't like to think (recently one of my students said in a seminar 'I don't like to have to think about things. I like clear, straightforward answers' – this is not a word of a lie). Therefore the combination of pressure and time and the more than occasional unresponsive student makes the type of participant teaching I like to do difficult. I also realise that it means that what I do is not always consistent with what I say and what I would like to do. This returns me to my pedestals. Those role models who are less than perfect have often been at the forefront of feminist geography. They have put up with all of this for years – in fact they have put up with worse because many of them did feminist geography when it wasn't as acceptable as it is today.

Many of us have jobs which are seen by some as the 'woman's job': the one who teaches something to do with gender (men apparently don't teach gender courses). The people who taught me, however, had to fight harder battles than me to get recognition for those courses. Frankly, I don't know how they had the staying power to do it but

I'm pleased that they did. In the first lecture I had from Nicky Gregson she introduced herself and gave a sort of potted history of her academic interests (to put her ideas in context). I was in my third year and no one had ever done this in my department before. I was embarrassed and I cringed. I even thought about changing courses afterwards. I now understand why she did this – cringey or not, it stayed with me. I remembered it on my first day of lecturing when I introduced myself. Consequently I was fortunate enough to be able to take the cringe factor out. To my introduction I added that I remembered someone else having done this before; it somehow legitimised my stance.

Clare Madge

An Ode to Geography

Geography,
What are you?
What makes you?
Whose knowledge do you represent?
Whose 'reality' do you reflect?

Geography,
You are not just space 'out there'
To be explored, mined, colonised.
You are also space 'in here'
The space within and between
That binds and defines and differentiates us as people.

Geography,
I want you to become a subject
On my terms and in my terms,
Delighting and exploring
The subtleties and inconsistencies
Of the world in which we live.

The world of pale moonlight and swaying trees in a bluebell wood.
The world of sand and bone and purple terror.
The world of bright lights flying past factory, iron and engine.
The world of jasmine scents and delicate breeze.
The world of subversion, ambiguity and resistance.
The world of head proud, shoulders defiant under the gaze of cold eyes laying bare the insecurity underlying prejudice.
The world of music, laughter and light,
Of torment and exploding violence
Of tar and steel strewn with hate
While the moon gently observes and heals.

Geography, could you be my world?
Will you ever have the words, concepts and theories

To encapsulate
The precarious, exhilarating, exquisite, unequal world in which we live?
I believe so.
By looking within and without, upside down and inside out,
Come alive geography, come alive!

Sarah Monk

I got into geography almost by accident. I had avoided it all my life, ostensibly because my mother was a geographer who despised it herself and became a town planner in her forties when forced to support herself and her children for the first time. I chose PPE but after a brief flirtation with industrial relations (seeking the workers' revolution perhaps? – I was in Paris in 1968) I found myself teaching land economics to surveyors, a course with a strong element of regional and urban economics. I then decided to try to become a 'real' economist and did a part-time evening MSc at Birkbeck but again chose urban and regional options. This got me a job in land economy which is also very spatial – and not 'real' economics anyway. After years of exploitation as a contract researcher, a PhD student who taught geography part-time at (then) CCAT (Cambridge College of Arts and Technology, now Anglia) convinced me to apply for their advertised post in economic geography; meanwhile he was busy convincing them that I would be ideally suited to the job. The match-making succeeded; I got the job and have been a geographer for nearly four years now. It has been a stimulating, mind broadening and above all an *interesting* experience. Not only did I find compatible colleagues but I also discovered that geography is a far wider subject area than economics ever was, even in its spatial guises. Further, out of a staff of ten, four are women, all feminists, and teaching on Anglia's MA in Women's Studies which I now help run. The opportunity to engage in feminist debates and to teach in a multi-disciplinary context has been a wonderful experience. I do not know which has changed my life more – geography or feminism!

I feel I have been trying to be a feminist all my life, from early tomboy days, to the sexual revolution of the sixties, a Women's Voice discussion group in the seventies, the Labour Party women's section in the eighties and now feminist geography. My main role model is probably Doreen Massey, who I first met in her days at the Centre for Environmental Studies (my children were in their crêche) and I was delighted to discover that as a geographer she makes a bloody good economist. My other role model is my friend Irene Bruegel, now lecturing at South Bank – I know no-one else with quite her energy and enthusiasm for the struggle. My 'burning question' is still one I had when I was an economist – why people's life chances are so different according to where they live and try to make a living – only today the focus is more clearly on women's life chances as I think they can only be explained in relation to the (male) social order(s) of the places where they live.

I am not sure if I manage to challenge or claim any categories or boundaries, yet I am still confounded by men's antagonism to me as a feminist. This often reduces me to silence or to sulky rejoinders – but why should I have to defend feminism to my line manager? Or indeed to any man?

Finally, I now feel sufficiently mature to recognise that I avoided geography at university not because it was such a second-rate subject, but because my mother got the best first in the whole of Rangoon University in ANY subject, which won her the only scholarship to UCL, where she again got the best first – I suspect that I believed I could never equal her achievement. Wanting her approval, I went to Oxbridge which she saw as superior – but I know the difference – and I only got a second!

Sarah Radcliffe

The categories 'feminist', 'woman', 'geographer' are so abstract and so intimidating when thought about in relation to myself – where do I position myself? Situating myself in relation to an ongoing wider intellectual narrative, in which I am implicated and represented in part, means to trace through my own (dis)connections with existing feminisms in geography.

Doing PhD fieldwork in the southern rural Andes of Peru, the agenda was one largely set by socialist feminist geography and by feminist geographers working in the geography of development. While both groups offered the promise of validity to my work (with peasant women migrating to cities around Peru), their inputs and influences were distinct. Socialist feminist geography called for research on work and reproduction in urban, industrial economies where issues of reproduction and production were mapped out so distinctly across space. Development feminists were faced with situations where the spatial and social patterning was varied and rarely fitted with 'Western' models. That was the theory that I grappled with in the field and in writing. In practice, what feminist geography (and other feminist writings) offered was an opportunity to bring feminist concerns, terminology, method and conceptual frameworks to an issue which I had chosen precisely for its focus on women (making women visible in academia also meant hopefully making them visible to planners and development programmes). I was thus trying, in a small way, to make women visible: feminist geography showed me the ways in which they had previously been made 'invisible' and then some ways for analysing their gendered experiences. Moreover, faced with two male supervisors whose knowledge of the feminist geographical literature was generally patchy, the existence of feminist geographical writing could be deployed critically during meetings to plan and then discuss my thesis research (my senior supervisor thanked me for introducing him to so much new literature!). In the early 1980s, the feminist literature in geography was diversifying rapidly, demonstrating a vitality and breadth of approaches which provided me, as an inexperienced research student, with a feeling of confidence that I was not tackling these questions on my own. While others may not have done research on 'my' topic, they were at least bringing critical insights into geographical approaches to work, household forms and migration, which I drew upon.

In addition to this initial link with feminist geography, my work has always drawn from a wider and largely interdisciplinary field of feminist theory (as is probably true for other feminist geographers). Given the unrelenting Euro- and North American-centrism of much geographical work (and unfortunately this includes much feminist work) during my time as a geographer, there have been times when I felt that feminist geography, for me, was more about invaluable social networks and feminist support for dealing with male-biased institutions and androcentric academic frameworks, than about 'feminist geographical theory'. However, given the on-going feminist concerns for critique and social change, and recent moves to recognise diversity and polyvocality in feminisms, my relationship to feminist geography is shifting once again. Still cognisant of the diversity of feminist voices – North and South – which inform practices and knowledges, feminist geographies now inform my research and teaching through their unprecedentedly strong and broad critiques of androcentric social and geographical theory. Perhaps I also feel that my research and my positionality (university staff, white, Western, a mother, etc.) requires that I now engage more directly in the debates and dilemmas which surround feminist geographies at the current time. From a position of non-innocence and by now quite lengthy involvement in geography, feminist geography is both 'out there' and 'the ground' of one dimension of my subjectivity: an ambivalent place to be, yet one which allows simultaneously for closeness and distance, critique and construct.

Gillian Rose

Like many of the other contributors here, I hesitated a long while before writing this testimony. I wrote it suddenly in a splurge of anxiety late in the course of this book's construction, in part as a response to the happiness and certainty that so many of the other contributions to this section seem to me to exude. For me, coming to feminism wasn't a revelation or a homecoming, it was a lifeline; there was no option but to take it because it gave me a way to make sense of my own uncertainty and confusion about my self. For a long while I felt like a paradox, which I often saw as a kind of fracturing gap inside me. As a person (never sure I was making it to being feminine), I felt hopeless, a failure because I was told I was. The only thing I was successful at apparently was academic work. I was bright at school, I won a place to Cambridge University, my initial stumblings around feminist historians' accounts of public and private space were encouraged by my Director of Studies, I won a studentship to do a PhD, three years later I got a permanent lectureship when many of my peers were on the dole or surviving precariously on short-term research contracts, and a few years later I published a book thanks to an initial contact set up by that same (ex-) Director of Studies of mine. I've had it relatively easy in the academy, helped by my Cambridge connections, the recent fashionableness of a certain kind of feminism, and my compulsive desire to be nice to people like me. But striating all that was what I felt just as a sometimes overwhelming anger, about why it was I did always perform so competently, so nicely. It was almost as if I had no choice, and that was somehow connected to the following rules, to even being produced by them as a person, a proper person this time (which again in the academy but ten times more problematically was, but might more properly not be – I was never quite sure – about being a woman), with just some angry residue left over which tainted it all. And feminism – theoretical, post-structuralist feminism – gave me a way to make sense of that paradox, of the performance and its angry excess. So my feminism feels intensely personal, and I worry that because it's so personal it might also become very individual. I still worry, my relationship to feminism still feels fraught, I worry about not being supportive and, ironically, about not being nice enough. But I think the important thing for me about feminist geography is that, through discussions and debates and relationships and readings, that gap inside no longer feels like a fault line, something that could rupture or overwhelm. Now I think of it less inside me and more like a space through which something new might emerge, and theoretically (which is bodily, emotionally) I'm more and more interested in what kind of space it might therefore be.

Jacky Tivers

In 1976 I completed a MSc degree in Geography at LSE (London School of Economics) and was offered a postgraduate research award at Kings College London to work towards a PhD. Having previously worked as a school teacher in England, and latterly as a university tutor in Australia, I was by this time in my mid-thirties, with a son of 18 months.

My initial 'strainings' towards a research topic led me in the direction of recreation and leisure behaviour. On one later-to-be-considered-momentous day my research supervisor suggested to me that I should abstract from the broad field of leisure studies the particular experiences and interests of one population sub-group. During discussion the idea was formed of focusing on women with young children. Clearly, this was a group with which I was readily able to empathise, since I was myself a member of it.

I went away from that tutorial discussion and read everything I could lay my hands on which might possibly be of relevance, starting with literature on leisure and then moving on to more general work about women. It didn't take long; there wasn't much to read! Very quickly I established that nothing had been written by geographers in Britain. There was a small and relatively inaccessible literature, concerning the Geography of Women, available in the US and Canada, dating from the very recent past. Most of this consisted of seminar notes and unpublished departmental papers. British literature sources were available in both Sociology and Psychology but, again, these were generally very recent and tended to focus on women as a total group, rather than women with children. Only one previous study (Gavron, 1966) had focused specifically on women with young children. It was hard to believe that so little research interest had been accorded to the problems, needs and behaviour of such a numerically and socially important population group.

The fact that I was working in a 'new area' for Geography was quickly picked up by the geographical 'establishment' and I was asked by the Urban Geography Study Group to present a paper, based on my research (entitled 'Constraints on spatial activity patterns: the case of women'), at the 1978 Annual Conference of the IBG in Hull. My approach was to call for the explicit consideration of women in geographical studies and the removal of their 'invisibility', through taking into account the existence of gender role differentiation in society. My 'surprise' at the way in which my paper was received has been described elsewhere (Tivers, 1981) (quote from a male colleague: 'I don't know what you mean by saying that geographers never look at women: we do it all the time').

I returned from the IBG Conference to the postgraduate room at Kings where, the following Monday, I wrote a paper for *Area* (Tivers, 1978a), before my sense of indignation had time to cool. It was deliberately polemical and I was a little afraid that it would be 'laughed out' by the Editor, who was, needless to say, male. I was indeed surprised and gratified that he printed the article so quickly (albeit with the subtle change of text from 'gender' to 'sex' throughout – an indication of the lack of understanding of feminist issues at that time). Almost immediately I also wrote an article for the USG (Union of Socialist Geographers) Newsletter (Tivers, 1978b) which was printed later that year.

I was now a Committee member of the Social Geography Study Group and volunteered to convene a session for the next Annual Conference on 'Family roles and spatial structure', which was to be the first ever IBG session specifically concerned with gender. By this time I was in touch with Sophie Bowlby and Laurie Pickup at Reading, and with Linda McDowell and Doreen Massey, both working in London. Late 1978 also saw the arrival in England of Suzanne Mackenzie from Canada. She found me at the office where I was working in London and rejoiced with me that feminist geography had at last been born in this country.

The session at the 1979 conference at Manchester was not an unqualified success! Suzanne and I gave papers but our visiting lecturer from the US was unable to attend and Sophie had to read her paper. The debate tended to descend into acrimony, with tight corners being fought by the Marxists who saw feminism as a detraction from the 'true cause' of socialism.

The rest, as they say, is history (or her-story). The establishment of a Women and Geography Working Party in 1980 was followed by the founding of the Women and Geography Study Group in 1982 (I was not at the Annual Conferences during those years due to the arrival of two more children – and therefore never quite counted as a 'founding member'!). By this time the number of people involved had grown considerably. I completed my PhD (Tivers, 1982) in September 1982 and was asked to submit it for publication soon afterwards (Tivers, 1985). The new Study Group agreed to write a textbook on feminist geography (WGSG, 1984) as a collective enterprise, and editorial responsibility was given to Sophie Bowlby, Linda McDowell (now at the Open University) and me. It was an amazing enterprise, involving hours on the telephone between Surrey, Reading and Milton Keynes, quite apart from all the work contributed by all the authors. This was followed by *Women in Cities*, which for me meant a 5000-word precis of my thesis (Tivers, 1988).

Nearly two decades after my first stumbling attempts to confront the 'maleness' of British Geography, it is unusual to attend any session at an IBG conference at which the word 'gender' is not mentioned. In 1976 there were about 40 women attending the conference, out of about 600 delegates. At Northumbria in 1995 there were 846 registered conference members, of whom about one-third were women, the Women and Geography Study Group ran well-attended sessions and many of the conference speakers were women. Women are less invisible in Geography, both numerically and in terms of subject content. There have been great changes, but there is still much to change, both in Geography and in the world which it seeks to describe and explain. I am proud to have been a part of the beginnings of feminist geography, and am proud too of the movement which has grown and expanded beyond the wildest dreams of the early days.

References

Gavron, H. 1966. *The Captive Wife*. London: Routledge.

Tivers, J. 1978a. How the other half lives: the geographical study of women. *Area*, 10, 302–306.

Tivers, J. 1978b. Yet another view of the Geography of women. *USG Newsletter*, 4: 25–26

Tivers, J. 1981. Perspective on feminism and Geography. In *Perspectives on Feminism and Geography*: papers presented at a meeting of the IBG Women and Geography Working Party, 26 September 1981, pp. 4–7.

Tivers, J. 1982. *Weekday spatial activity patterns of women with young children*. University of London.

Tivers, J. 1985. *Women Attached*. London: Croom Helm.

Tivers, J. 1988. Women with young children: constraints on activities in the urban environment. In Little, J., Peake, L. and Richardson, P. (eds), *Women in Cities: Gender and the Urban Environment*. London: Macmillan.

WGSG. 1984. *Geography and Gender*. London: Hutchinson.

Jenny Williams

I came into academe as a mature student, single mother of two small children, having enrolled for a BA in Social Studies at the local poly. It wasn't until I'd completed my foundation year that I decided to specialise in geography rather than sociology – if the truth be known, because the geographers offered fieldwork in Africa and Amsterdam!

Before that, during my career in the Civil Service, I'd experienced dissatisfaction in terms of both lack of career opportunities for women, and the unfairness of the legislative policies which I was paid to enforce, which I found to be discriminatory against workers in general and women in particular. The first of a large family to 'do a degree', I expected some of these (as yet unnamed) uncertainties to be properly explained.

I was taught geography as an objective discipline (by men) from textbooks reflecting the interests of, and written by, men. My reality was not incorporated into such materials; my spatial experiences were treated as inconsequential; I felt excluded.

Why, I asked tutors, in the late 1980s, were my continual pleas to recognise gender differences over use of space always dismissed as the whining of a single feminist?

Shortly before I finished my third-year dissertation, which attempted to develop a feminist geographic analysis of the refurbishment of an inner city housing estate, I discovered a book on the library shelves called *Geography and Gender*. Serendipity! Having been taught from a masculinist standpoint (with the exception of one radical lecturer who was shunned by his colleagues) I was stunned to find published work by women who shared my sense of exclusion. For my undergraduate studies, of course, this was too late – but for my own intellectual development, this was a day of revelation!

I subsequently joined the WGSG, and have been active writing, teaching and organising around feminist politics ever since. I also acted as secretary in the Institute of British Geographers Equal Opportunities Working Party in an attempt to be proactive rather than passive about the ongoing inequalities in geography. Teaching in two different universities, both times as the sole female member of academic staff, I often still feel the exclusion created by the patriarchal profession I work in. However, the battle is worth it, and several more feminists graduate from one Northern university than they did five years ago!

Reflections

Three of the major problems which autobiographical material poses for any researcher are how to represent such material; the extent to which such material should be subjected to further analysis; and how any such analysis might be done: see, for example, Janet Townsend's discussion of these issues in *Women's Voices from the Rainforest* (Townsend, 1995). Those of us who have been involved in writing this book differ considerably in the ways in which we use this type of material in our own research (compare, for example, Townsend, 1995; Gregson and Lowe, 1994; Valentine, 1993a). Here, however, we have been concerned to use these autobiographical testimonies to illustrate how different people write different histories and to demonstrate the complexity of contextual histories. Reproducing the testimonies verbatim therefore has been a strategy which we've agreed on. Where we differ, though, is in the extent to which we, writing as individual researchers, would comment on the testimonies. For some of us, the desire to use testimonies as examples of situated knowledges precludes drawing out universal conclusions from the testimonies. Others of us, particularly those for whom the pedagogic role of this text has been of paramount concern, have argued that, given this purpose, we cannot simply leave the testimonies as they stand, uninterpreted. At the same time, though, those of us who have argued in this way have taken on board the problems which would be posed by the reintroduction of a universalising discussion. The compromise position therefore is to highlight a series of issues, but to do so in terms of some of the differences and tensions which we feel are raised within various of the testimonies. We ask you to reflect on these issues yourselves.

ACTIVITY

Re-read the testimonies and, as you do so, think about the following questions and issues.

Feminism and feminist geography. Not surprisingly, several of the testimonies reflect on this. Feminism is revealed as meaning different things to different people; so too is feminist geography, and geography. Clearly different people have had differing relations to feminist geography, and this has varied over time, even in relation to the same individual. Try to outline the nature of these differences. What reasons can you suggest for these? In thinking about this second question it might be helpful to think about what the testimony writers choose to say about their own positionality.

The importance of institutional contexts. Many of the testimonies comment on how institutional contexts, in the form of particular departments and informal networks, have been influential to them. Try to outline the various ways in which these influences work. Do particular institutions appear to be more or less conducive to the development and teaching of feminist geography, and to feminist geographers' research? Why might this be? What relationship exists between informal networks and departmental contexts? How supportive do you think your own institutional context is to the development of feminist perspectives?

Personal politics and personal experiences. One of the key arguments made by the feminist movement in the 1970s was that 'the personal is political'. Women's everyday and personal experiences were seen as raising issues which needed to transcend their location in the private, invisible worlds of individual households. They needed to be politicised. In contrast, such arguments are frequently criticised today for their essentialism and their uncomplicated acceptance of the authenticity of women's experiences. However, little has been suggested to take their place and many feminists criticise the latter position as one which encompasses 'a politics of nowhere'. Reading the above testimonies, how influential do personal experiences appear to have been for feminist geographers? How important have personal experiences been in shaping your own politics? How do you respond to the arguments which question the authenticity of women's experiences? Do any of these testimonies appear to be a 'view from nowhere'? You might like to reflect on your answer here. If you answered 'yes' to this last question, what made you think this? Contrastingly, if you answered 'no', what might this say about the politics behind testimonial writing? If you were undecided about labelling particular testimonies thus, why do you think this was?

Role models. A number of the testimonies make specific reference to various individuals (mothers, friends, academics – both friends and colleagues) who have played a major influence (knowingly or unknowingly) in their lives. Consider the various ways in which the testimonies show role models to operate. Why should role models be so important to some feminist geographers? Have role models been important to you, and if so how and why?

Writing autobiographically. The testimonies are written using various approaches and writing styles, some entirely personal, some more conventionally 'academic'. Consider the range of approaches appearing here and explore why you think particular people choose to write and represent themselves in particular ways. Finally, we'd like you to write your own testimony, exploring your own intellectual or academic development. This may appear an easy exercise at first, but we're sure that you'll find it as difficult as we have to put into practice!

SUMMARY

Feminist geography's history can be written as a multiply voiced narrative, to emphasise the heterogeneity of the feminist geographic tradition, its negotiated and contested nature and its complex contextuality.

2.5   Summary discussion

In this chapter we have focused on the diverse and different ways in which feminist geographers have chosen to write the history of feminist geography. From this we can see that there is no one history of feminist geography, rather several histories, and all are written for particular strategic purposes: to demonstrate intellectual maturity, to show connections with the broad base of feminist scholarship, to recover one or more different (feminist) traditions of Geography, to demonstrate the importance of situated knowledges. Writing history, then, is an intensely contemporary process, in which the differences between us are significant to understanding the histories which we choose to write. We would like you to come away from this chapter, therefore, with an understanding of each of these ways of writing the history of feminist geography, and of the differences between them, and with an appreciation of the subtleties of our own position here. What we are categorically not arguing for in this chapter is that any one of these positions is inherently better than any other, although admittedly we all have our personal preferences. Indeed, the last thing which we want you to infer from the chapter is that the order in which we present things here represents a sequentially 'better' way of writing 'the history' of feminist geography. Rather, we maintain that we need to continue to produce multiple histories for multiple audiences. Thus, we need to continue to produce histories which represent progress and intellectual maturity, if only to engage with the dominant discourse of knowledge as progress. We need to construct feminist traditions of Geography to demonstrate the persistent erasure of women from the geographical tradition. And we need to produce multiply voiced histories to show that – like all geographical traditions – feminist geography is heterogeneous; that it is 'situated messiness', contested and negotiated by particular bodies in particular places. Just as (a few) male historians of geography are prepared to admit their uncertainties within the geographical traditions they construct, so feminist geographers too need to show that we also are '[n]either as singular, [n]or as confident as we, in our singularly confident moments, are tempted to presume' (Driver, 1995: 413). In short, all of these ways of writing the history of feminist geography (and almost certainly more) matter, and need to be continued with. In so doing, not only do we demonstrate the heterogeneity of feminist geographical knowledges but we also destabilise mainstream homogenising representations of the geographical tradition which all too frequently erase feminist knowledges from their script.

READING A

Stoddart, D. R. 1991. Do we need a feminist historiography of geography – and if we do, what should it be? *Transactions of the Institute of British Geographers*, 16, 484–487.

Domosh (1991a) attempts to make a case for a feminist historiography of geography in the context of what she terms 'the post modern turn', and she particularly links this attempt to the presumptive deficiencies in this regard of *On Geography and its History* (Stoddart, 1986). I find her reading of this book misleading and her proposal unconvincing.

  First, the book is in no sense a history of geography... It is a commentary on the emergence of an academic discipline dominated by a particular natural science tradition... Domosh finds it odd that various nineteenth-century women travellers 'are not even in Stoddart's book': she comments that I am not alone 'in writing a man's story of geography, but by celebrating the exploratory tradition in geography, his omission of women is even more blatant than many other authors'...

The simple fact is that none of the persons Domosh discusses have anything to do with my themes: there is therefore no reason to mention them. Ms Domosh is doubtless unhappy with this, but it accurately reflects the historical reality of the academic development of the subject with which I was concerned...

Domosh argues that a group of Victorian women travellers made a specific and distinctive contribution to the emergence of the discipline (or, she appears to say, perhaps they did not). She is able to maintain her argument for only two reasons. The first is loose use of language: 'travel' is continually equated with 'exploration', and 'exploration' with 'fieldwork'...

Second...we have only the remarkably lame assertion that 'With the hindsight offered by historical reflection, we can point to the potential contributions of Victorian women travellers'... Potential? Can it therefore be surprising if these unrealised achievements made no impact on a rapidly evolving discipline?...

Domosh's agenda is revealed in her conclusions: 'Geography's commitment to a value-free, perspectiveless, objective science must be questioned in light of feminist critiques'...such a posture simply imposes anachronistic interpretations on what was happening a century ago, when geography was indeed...dominated by 'white, male, aristocratic heroes'... It would indeed be historiographically pointless either to deny this or to condemn it...

[He then goes on to comment on the work of a few female geographers, notably Ellen Churchill Semple, within the discipline.]

Domosh, M. 1991b. Beyond the frontiers of geographical knowledge. *Transactions of the Institute of British Geographers*, 16, 488–490.

I salute Mr Stoddart for taking to heart my criticism of his book *On Geography* for its failure to include women as significant figures in geography's exploratory heritage... According to him, the women whom I discuss in my article have nothing to do with the themes of his book, and therefore are not relevant to his arguments or discussions. Furthermore, Stoddart is contending that this omission of women does not reflect any sexism on his part, but rather 'accurately reflects the historical reality'. I agree with him here in regard to the latter half of that sentence. Women have not figured prominently in the history of geography, and it was that very fact that prompted me to write this article. My question was why – why, in the historical reality we have created, women and women's experiences only appear as exceptions to the 'regular' course of history?

One of the reasons that I chose to focus my discussion on these Victorian 'travellers' was because their lives and stories coincided with the establishment of geography as a scientific discipline. What I questioned was the basis of the 'emerging standards of the time' that served to disqualify women 'travellers' as geographers...

...Francis Younghusband 'made no systematic observations at all' on his trip through Asia...although he was granted the RGS's (Royal Geographical Society) gold medal. I doubt that Isabella Bird's observations on her varied trips were any less systematic. Mary Kingsley considered herself part of the English scientific community, and she undertook planned trips to West Africa some of which could be called, in Stoddart's terms exploratory...and some of which improved fieldwork in the sense of the systematic collection of wildlife specimens...and cultural artefacts. However, when she reached the summit of Mount Cameroon, having lost all her male companions along the way, she did indeed leave her calling card, but the initials FRGS (Fellow of the Royal Geographical Society) were *not* next to her name...

...most of the founders of 'modern' geography, and those who chronicle their accomplishments, have not been able to accept women as active 'negotiators' in geography, nor have they been able to accept a vision of geography different than their own...

READING B

Pratt, M. L. 1992. *Imperial Eyes: Travel Writing and Transculturation.* London: Routledge, pp. 214–215.

...

The Lady in the Swamp
It is hard to think of a trope more decisively gendered than the monarch-of-all-I-survey scene. Explorer-man paints/possesses newly unveiled landscape-woman. But of course there were explorer-women, like Alexandra Tinne and Mary Kingsley, who led expeditions in Africa, and explorer-wives like Florence Baker, who accompanied expeditions up the Nile...these women, in their writings, do not spend a lot of time on promontories. Nor are they entitled to. The masculine heroic discourse of discovery is not readily available to women... Mary Kingsley's extraordinary *Travels in West Africa* (1897) is probably the most extensive instance that does exist. Through irony and inversion, she builds her own meaning-making apparatus out of the raw materials of the monarchic male discourse of domination and intervention. The result, as I will suggest below, is a monarchic female voice that asserts its own kind of mastery even as it denies domination and parodies power.

Kingsley went to West Africa around the age of 30 as an entomologist and ichthyologist mainly interested...in the small-scale life forms that inhabit the vast and unexplored mangrove swamps of the Gabon. The domain she chose to occupy, then, could hardly contrast more starkly with the gleaming promontories her fellow Victorians sought out. Indeed, 'her' swamps, as she calls them, are a landscape that the Africans themselves seem neither to use nor to value, a place where they would never contest the European presence. Kingsley depicts herself discovering her swamps not by looking down at them or even walking around them, but by sloshing zestfully through them in a boat or up to her neck in water and slime, swathed in thick skirts and wearing her boots continuously for weeks on end. Her comic and self-ironic persona indelibly impresses itself on any reader of her book. Here she is in a famous passage, fresh out of the interior and hitching a ride to the coast in a small boat with a blanket for a sail, as usual the only European and the only woman in the party:

> As much as I have enjoyed life in Africa, I do not think I ever enjoyed it to the full as I did on those nights dropping down the Rembwe. The great, black, winding river with a pathway in its midst of frosted silver where the moonlight struck it; on each side the ink-black mangrove walls, and above them the band of stars and moonlit heavens that the walls of mangrove allowed one to see. Forward rose the form of our sail, idealised from bedsheetdom to glory; and the little red glow of our cooking fire gave a single note of warm colour to the cold light of the moon. Three or four times during the second night, while I was steering along by the south bank, I found the mangrove wall thinner, and standing up, looked through the network of their roots and stems on to what seemed like plains, acres upon acres in extent, of polished silver – more specimens of those awful slime lagoons, one of which, before we reached Ndorke had so very nearly collected me. I watched them, as we leisurely stole past, with a sort of fascination... Ah me! give me a West African river and a canoe for sheer good pleasure. Drawbacks, you say? Well, yes, but where are there not drawbacks? The only drawbacks on those Rembwe nights were the series of horrid frights I got by steering on to tree shadows and thinking they were mudbanks, or trees themselves, so black and solid did they seem. I never roused the watch, fortunately, but got her off the shadow gallantly singlehandedly every time, and called myself a fool instead of getting called one... By daylight the Rembwe scenery was certainly not so lovely, and might be slept through without a pang.

What world could be more feminised? There shines the moon lighting the way; the boat a combination bedroom and kitchen; Kingsley the domestic goddess keeping watch and savouring the solitude of her night vigil. Far from sharing her joy, the party, thank goodness, are asleep. The place is almost subterranean – like a mole, the traveller peers through roots and stems. Beauty and density of meaning lie not in the variety and colour that unveil themselves, but in the idealisation which the veil of night makes possible *in the mind of the seer*. By day, one sees not variety and density, but their opposite, monotony. Which is to say that Kingsley creates value by decisively and rather fiercely rejecting the textual mechanisms that created value in the discourse of her male predecessors: fantasies of dominance and possession, painting that is simultaneously a material inventory. She foregrounds the workings of her (European and female) subjectivity: the polished silver is the product of her own imagination at work on a mangrove swamp. Far from taking possession of what she sees, she *steals* past; far from imagining a civilising or beautifying intervention, she contemplates only the silly possibility of 'damaging Africa' in a collision that would doubtless damage her worse...

READING C

Blunt, A. 1994. Reading authorship and authority: reading Mary Kingsley's landscape descriptions. In Blunt, A. and Rose, G. (eds), *Writing, Women and Space: Colonial and Postcolonial Geographies*. London: Guilford Press, pp. 63–67.

... It is persuasive to locate Mary Kingsley in a 'swampy world' – part of but separate from imperialism – but this location reduces the ambivalence of her identity. Mary Kingsley herself used a topographical metaphor to view Africans inhabiting swamps far below the peaks of Western 'civilisation':

> I do not believe that the white race will ever drag the black up their own particular summit in the mountain range of civilisation...alas! for the energetic reformer – the African is not keen on mountaineering in the civilisation range. He prefers remaining down below and being comfortable. He is not conceited about this; he admires the higher culture very much, and the people who inconvenience themselves by going in for it – but do it himself? No. And if he is dragged up into the higher regions of a self-abnegatory religion, six times in ten he falls back damaged, a morally maimed man, into his old swampy country fashion valley.

This quotation raises questions about Mary Kingsley's own position as a white woman travelling through literally and metaphorically colonised landscapes...the spatiality of ambivalent subjectivity should not...be confined to literal and metaphorical swamps. For example, Mary Kingsley's ascent of Mount Cameroon potentially implicates her in masculine and imperial discourses, but, by ambivalently doing so, reveals the complexities and contradictions of her place within such discourses... Mary Kingsley's account of her ascent illustrates the ambiguities of being constructed as both 'inside' and 'outside' and moving between patriarchal and imperial discourses. Her description of her ascent lacks the lively enthusiasm and good humour of the rest of *Travels in West Africa*. For example, her ability to identify with the masculine, imperial trope of panoramic vision is undermined when her view is obscured by mist. Furthermore, this obscured vision seems to enhance the natural beauty of the scene in her eyes, as she perceives it in aesthetic rather than strategic terms:

> The white, gauze-like mist comes down the upper mountain towards us: creeping, twining round and streaming through the moss-covered tree columns... Soon...all the mist streams coalesce and make the atmosphere all their own, wrapping us round in a clammy, chill embrace; it is not that wool-blanket, smothering affair that we were wrapped in down by Buana, but exquisitely delicate. The difference it makes to the beauty of the forest is just the same difference you would get if you put a delicate veil over a pretty woman's face or a sack over her head. In fact, the mist here was exceedingly becoming to the forest's beauty.

It is particularly interesting that Mary Kingsley here employs an objectifying, masculine metaphor. The landscape is feminised but its attraction lies in veiling rather than unveiling. From the peak, however, Mary Kingsley's view is fully obscured by thicker, less aesthetically appealing mist. This symbolically reflects her own position as attempting but unable fully to achieve masculine and imperial vision... Mary Kingsley's ascent of Mount Cameroon ironically and ambivalently locates her both inside and outside a masculine, imperialist tradition of exploration, conquest and surveillance, illustrating the complexities and contradictions of subject positionality... Rather than counter Mary Louise Pratt's siting of Mary Kingsley in swamps by simply describing her on a mountain, I hope that by stressing the spatiality of ambivalent subject positionality, fixity and movement are kept in tension instead of in isolation.

READING D

Gould, P. 1994. Guest essay/essai sur invitation: sharing a tradition – geographies from the enlightenment. *The Canadian Geographer*, 38(3), 194–200.

...

Radical Feminist Geography

Given this...context of questioning, I want to stand in that tradition of critique and challenge some of the challengers. First, feminist geography, where I am sure a questioning and critical statement from one of Geographia's many lovers will be greeted with less than enthusiasm. I am unaware of any critical statements in print of radical feminist geography, which may mean it has been elevated rapidly to the sacred, where faith and belief rule, but reason is not relevant...nothing I say should be construed as a devaluation of thoughtful feminist critique and the perspective it provides as yet another valuable condition of possibility for seeing. A 'world' of women in a small Italian–Swiss valley, and an imaginative historical 'reconstruction' and speculation, are sensitively illuminated by Verena Meier; the spatio-temporal constraints on women in a small Swedish town are highlighted by Solveig Martensson; Doreen Massey focuses explicitly on the effects of economic restructuring on women and stands in the critical tradition herself; while Julie Graham graces geography with her thoughtful challenges to a stultified Marxism.

But under the exploding hegemonic power of strident feminist critique, we see increasingly male geographers scurrying around in print, flashing their gendered sensitivity. Their precious posturing, informed by a gloomy morality, obliterates even an historical understanding of the very language in which they write. They would rather destroy the potential for rhythmic structure in an English sentence, with an ugly 'she/he' or 'his/her', or even 's/he' than use one or another nominative or possessive form in their interchangeable possibilities. Possessive adjectives are not a mark of phallocentrism; they arise in the structure of some languages and not others...As for the constant interjection of 'sic'...I find the posturing so shallowly flamboyant and historically uninformed and ignorant, that it may fairly be called 'sic-ening'....

But there is a deeper concern here. For the first time in our literature, I have come across not just anger, which is perfectly understandable, but something close to hate. And I object. Why should my very being, and the 'world' into which I was thrown, and about which I had no choice, be used epithetically by clitorocentrists of any gender? I *am* a 'white European male'. It is true, I am not dead yet – but give me a chance....feminist epithets are not needed in any caring human discourse, and I suggest that it is time we all grow up...

Peake, L. 1994. Proper words in proper places... or of young turks and old turkeys. *The Canadian Geographer,* 38(3), 204-206.

At the last annual meeting of the Canadian Association of Geographers in Ottawa, I attended the Wiley Lecture given by Professor Peter Gould. I was joined by a few other feminist geographers, but this was hardly surprising given the furore that greeted the publication of his book, in 1985, entitled *The Geographer at Work.* Professor Gould has a reputation for being less than sensitive to the arguments put forward by many feminist geographers about the inherently masculinist bias of our discipline. Unwilling to believe, however, that a leopard cannot change its spots and wanting to hear his views at first hand, I sat down to listen. What follows are my thoughts on sharing a tradition....

What I found most extraordinary was the insidious manner in which the audience was invited to share his vision of Geographia, depicted in *The Geographer at Work* as a naked woman. It was an invitation that could be understood only on the assumption of shared interests in the representation of geography in a particularly prurient manner. And it is not only that Professor Gould was displaying a disturbing level of sexism in choosing to ignore the extent to which other geographers had originally found the illustrations offensive and distasteful. It is what these illustrations tell me, not only about Peter Gould...but more importantly about definitions of masculinity and of the relationship between sexuality and modernism. ... If references to such material raise for me such questions as how women are being represented and from whose point of view and what are the political effects of such icons, then why aren't the same questions being raised for somebody (else)...

It is not that I wish to engage in author bashing in this commentary. There are far more important issues at stake... But his words ring false when he slips so often from critique of the object of his analysis to criticism of the subject...the section which (is) ostensibly a critique of radical feminist geography becomes a vitriolic attack on radical feminist geographers who do not conform to his notions of fair academic standards...

...The dominant image he presents of himself ...is that of a profoundly patriachal modern man...(un)aware of his own dominant position as a white, Western male and of the association of this with the universal. He is somewhat willing to let in the voices of those previously excluded, but woe betide any feminist who adopts a mode of expression that offends his notion of clearness, effectiveness and beauty. 'Fine scholars' make 'fine prose' it would seem by 'gently' and 'calmly' arguing their position.

But this situation is fine only if we accept the male prerogative to legislate language... The problem ...is not one of women writing hatefully; the problem is the notion of a single universal subject, which has been constituted in the form of white, Western man. It is this male symbolic order which is under attack in women's writings... And Professor Gould is reacting to this attack on a personal level...

Professor Gould suggests that it is time we all grew up. The problem is that Young Turks, the (exclusively male) intellectual vanguard of academic geography...have a habit of growing up into Old Turkeys. Maybe it is time not only to grow up, but for some Old Turkeys to come down off their roosts. That would be just fine by me.

Gould, P. 1994. Reply, *The Canadian Geographer*, 38(3), 209–214.

In these post modern days of different perspectives and hermeneutic stances, anyone who thinks that a publication, a 'making public', will come to reflect the author's intentions is either ignorant or naive... So I thank my commentators...for the courtesy of their responding. Two of them, however, do not make it easy to respond in turn, since their commentaries barely touch, let alone engage, the questions that the text provided...

It is equally difficult to respond to what can only be termed the tirade of Linda Peake, who makes it clear that she had formed her opinion before even walking into the lecture hall. She did not like my book *The Geographer at Work*... the 'furore' came not from the thousands of readers, women and men, who found the depictions of the goddess Geographia inoffensive, amusing and symbolic, but a handful of radical feminists who had formed themselves into the Geographical Perspectives on Women Speciality Group...the first two depictions of Geographia were...comments on the abduction of our goddess by uncouth quantifiers and Marxists. The third, ...showed a complete reversal, a beautiful Geographia fully in command, carrying a bunch of men in total disarray back to Geographical Reality...anyone whose sense of humour has not been totally lobotomised sees this as symbolic of the rise of the feminist perspective in our field...

...I suggest it is time that some people grew up.

CHAPTER THREE

Gender in feminist geography

NICKY GREGSON, UMA KOTHARI, JULIA CREAM, CLAIRE DWYER,
SARAH HOLLOWAY, AVRIL MADDRELL AND GILLIAN ROSE

3.1 Introduction

Our aim in this chapter is to introduce some of the central ideas shaping how feminist geographers think about the world they study. As such, the focus here is inevitably theoretical, and this in itself we feel worthy of a few preliminary comments. Being university teachers, we all know the mere mention of the terms 'theoretical' and/or 'theory' – at least in a British context – is likely to instil negative reactions amongst students. 'Theory' for many British students seems readily labelled as 'hard' and/or irrelevant, an invitation to switch off rather than switch on. In this chapter, though, we set out to show otherwise: yes, theory requires you to think but this thinking is exciting, stimulating and full of possibilities. Furthermore, and as we show throughout this chapter, as well as enabling you to interpret empirical realities, theory is critical to understanding why researchers approach particular topics in particular ways.

For feminist geographers, the central analytical category, at least until the mid-1990s, has been 'gender'. In this chapter, therefore, our main objectives are to answer the question 'what is gender?' and to show how feminist geographers have worked with this concept.

3.2 What is gender?

Drawing broadly on feminist work throughout the social sciences, feminist geographers have for the most part argued that gender is a social construction which draws on certain aspects of biological sex. Sex itself is assumed to be a natural category, based on biological difference.

ACTIVITY

The following is a list of statements about men and women. Which are statements about sex and which are statements about gender?

- Women give birth to babies, men don't.
- Little girls are gentle, boys are tough.

- In one case when a child brought up as a girl learned that he was actually a boy, his school marks improved dramatically.
- Amongst Indian agricultural workers, women are paid 40–60% of the male wage.
- Women can breastfeed babies, men can bottlefeed babies.
- Most building-site workers in Britain are men.
- In Ancient Egypt men stayed at home and did weaving. Women handled family business. Women inherited property and men did not.
- Men's voices break at puberty, women's do not.
- In one study of 224 cultures, there were five in which men did all the cooking, and 36 in which women did all the housebuilding.
- According to UN statistics, women do 67% of the world's work, yet their earnings for it amount to only 10% of the world's income.

Source: *The Oxfam Gender Training Manual* (S. Williams, 1994).

From the moment they are born (and sometimes before), human beings are treated differently because of their sex. In contemporary Western society, for example, one of the first questions people ask of a newborn baby is 'is it a boy or a girl?', and a look at any baby care shop will show you how clothes for newborns are designed to answer this question visually when the immediate physical appearance of the child may be more indeterminate. The following extract exemplifies this imperative to classify.

Avril Maddrell

The case of baby Sam

I have been amazed at some reactions to my baby Sam. Initially born with a full head of dark hair and well-rounded features, he was on several occasions described not merely as pretty, but specifically 'too pretty to be a boy'. The fact that as parents we chose not to dress Samuel in chunky 'boy's' style clothing from birth, combined with his looks, clearly left some people puzzled as to how to identify his sex. 'What a pretty girl'; 'Oh I thought he was a girl because he's so pretty' were two of the common remarks people made.

For some people, ideas of physical beauty are so strongly associated with gender that they lead to assumptions about sex. In Samuel's case, his hair and features were identified by some as feminine, leading to the assumption that his sex was female. The fact that attractiveness is in the eye of the beholder (i.e. a cultural construction) and not the prerogative of the female sex only serves to underline the extent to which gender stereotypes inform our social interaction.

ACTIVITY

Look at each of the photographs in Figures 3.1–3.4 and decide whether they are of a boy or a girl. On what attributes or characteristics do you base your decision?

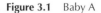

Figure 3.1   Baby A

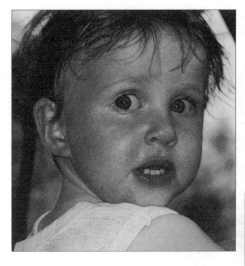

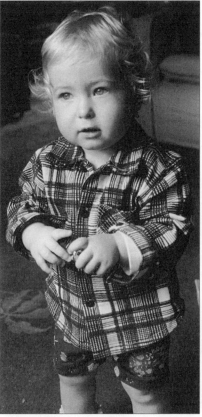

Figure 3.2   Baby B

The differential treatment of males and females continues throughout their lives, from the toys boys and girls are given to play with to the jobs that are considered appropriate for men and women. The male and female sex are therefore *gendered by society* as boys and girls, men and women. Boys and men are expected to exhibit masculine characteristics. As children they are expected to play with cars, as adolescents they are expected to be boisterous, and as men they are assumed to take responsibility for a situation and lead.

Figure 3.3   Baby C

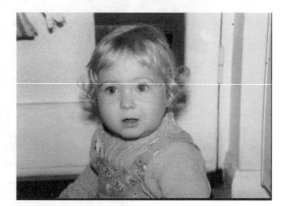

Figure 3.4   Baby D

Girls and women are expected to exhibit feminine characteristics. As children they are expected to play with dolls, as adolescents they are expected to be compliant, as women they are expected to care and follow. And there are sanctions for boys and girls, men and women who don't conform to the masculine and feminine characteristics attributed to them, for those who don't perform the correct gender for their sex. For example, boys who play with dolls may be ridiculed (unless the doll wears military clothing) and women who don't want to be mothers, who don't want to take on their traditional caring role, are seen as unnatural.

When we look at these characteristics we can see that they are defined in relation to one another. They work as pairs of opposites: when boys play with cars, girls play with dolls; where boys are boisterous, girls are compliant; when men lead, women are meant to follow. This oppositional relation is particularly important because each side of the pair is not equally valued. In simple terms, masculine characteristics are valued more than feminine ones; people who lead are valued more than people who follow. Thus, it is not just the case that males and females are gendered differently; rather, they are gendered differently and as a result valued differently. The social constructions of the two genders relate in a way that works to the general advantage of men and to the general disadvantage of women.

SUMMARY

- Sex is a natural category based on biological difference.
- Gender is a social construction organised around biological sex. Individuals are born male or female but they acquire over time a gender identity, that is, what it means to be male or female.

3.3   How do feminist geographers work with gender?

In the previous section we considered the question 'what is gender?' In this section we shift the focus somewhat, to consider the various ways in which feminist geographers work, and have worked with the concept of gender. At this stage our concern is to keep things relatively straightforward. We therefore present feminist geographers' ways of working with the concept gender as a chronological narrative. However, we do not want you to infer from this that the various interpretations presented here represent categorically different and progressive stages in the history of feminist geography discourse. Rather, we emphasise that the positions outlined are much more fluid than they appear here, that they frequently overlap in time, and that individual feminist geographers themselves can hold more than one of these positions, sometimes simultaneously. At certain junctures in this section, therefore, and particularly in our discussion of various examples, we emphasise where individual authors seem to be thinking in terms of more than one interpretation of the category gender. This may be within the confines of one single piece of work, but it can just as easily occur when we take a historical perspective on the writings of individual feminist geographers. Two brief personal sketches illustrate what we mean here. Although at this point in the chapter you might find it difficult to grasp all that these writers are arguing, by the end of the chapter these personal trajectories will be clearer:

Janet Townsend

I came in by accident, in 1984. For nearly twenty years my research had been in Latin America, with people pulling down rainforests and trying to make farms and a future for themselves and their children. An amnesty for guerillas in Colombia gave me the chance to go back after a decade of exclusion to places which I knew well, and I wanted a cover story under which I could look at change in the political economy of pioneering. So I set out to look at women's roles on the frontier of settlement and to prove their importance to production, all as a cover. I found the pioneer women to be housewives and the gender relations a great deal more interesting than the political economy. I'd never asked pioneer women any personal questions, just stuck at the level of the household for eighteen years. This is a tale I've told before (Townsend, 1993; Townsend et al., 1995): pioneer women led me into feminism and into a new life. Before that, a male colleague had asked me, 'Why don't you go into women and geography, Janet? There's real resources for research'. And I'd replied, 'There are much bigger differences in the world than those between men and women'.

In 1984, the IBG [Institute of British Geographers] had come to Durham, and Janet Momsen and I had put together a session on 'Women's role in changing the face of the earth', which was all papers on (not from!) poorer countries. Mark Cohen of Hutchinson's liked it, and asked us to build it into a book, which became *Geography of Gender in the Third World* in 1987 (N.B. including caveats about the use of 'Third World') (Momsen and Townsend, 1987). Janet's contacts gave us the contributors, although we had to work hard to get it together. Then, there we were with the chapters, looking for a book. The chapters were neither innovative nor particularly feminist save in representing the very late arrival of geographers on the Women in Development bandwagon. The chapters had new geographical material on women (mostly not gender) in poorer countries, some of it very important, but did not represent a significant geographical development of the field. For any feminist geographer interested in gender in poorer countries, there were plenty of better things to read – but not with the word 'Geography' in the title. We were in a discipline where women, never mind about gender, were still invisible to most. So we struck an old-fashioned blow for the visibility of women, as if that invisibility had been an accident, and edited the book for male geographers and unawakened women, seeking to use it to get women into the curriculum of 'development geography'. We expected to be competed off the map in a couple of years, and were very upset when Hutchinsons took two years from receipt of the final text to publish it! It was not out of ignorance that we accepted the agenda of male geography in editing the book but a deliberate choice of what we saw as the most effective use of our material in a long revolution.

I wasn't in on *Geography and Gender* (WGSG, 1984) or on the heady early years of the Study Group, but I'm still amused by the similarity of group meetings now to the mid-80s. We may have sought to establish the field by conforming, but it wasn't that we didn't discuss radical issues, rather that we never thought anyone would publish what we had to say about them. Certainly, I never considered it; perhaps others did. As a geographer I felt pretty incapable of scholarly comment on male violence. Yes, we were very effectively silenced, and we silenced ourselves. The group was a haven for the discussion of male coercion, harassment, sexuality, whether the central problem is penetration and so forth, yet so far as I know no member published on these issues!

Dutch geographers opened my eyes to a central feature of geography. They invited members of the group to a week's meeting in Amsterdam, and kept asking us 'Why do you always write and talk in terms of the nuclear family?' We felt insulted, the nuclear family being an orthodox enemy, until we realised that the real criticism was of heterosexism. Why were none of us lesbian? Were there lesbian members of the group who had not come? I was deeply embarrassed to realise that I couldn't think of anyone out, and in fact that none of my lesbian or gay friends were geographers. What was wrong with us? I'd never noticed. I was used to geography as a colonial, imperialist discipline, but had never recognised how absolutely it was controlled by the hegemonic masculinity of the time. And I still didn't see how completely male is the gaze...

Feminisms in geography have come a long way, which warms my heart. It's fun to be involved. The only feature of present feminist geographies which I don't like is the development of elite languages which exclude others. I'm old-fashioned. I believe that we need to change, to capture, to subvert the hegemonic discourses, not to set up fascinating exchanges for a tiny minority! Oddly, the criticism that feminist geography in Britain is elitist has been going strong since its beginnings. I hope that I shall not see it justified.

References

Momsen, J. and Townsend, J. (eds). 1987. *Geography of Gender in the Third World.* London: Hutchinson.

Townsend, J. 1993. Gender studies: whose agenda? In Schuukman, F.J. (ed.), *Beyond the Impasse: New Directions in Development Theory.* London: Zed Press.

Townsend, J. *et al.* 1995. *Women's Voices from the Rainforest.* London: Routledge.

WGSG. 1984. *Geography and Gender.* London: Hutchinson.

Nicky Gregson

I first started thinking about feminism in geography shortly after the appearance of *Geography and Gender,* at a time (the mid-1980s) when grand social theory was the order of the day in human geography. This context is central to understanding the vision of 'gender' which I was then working with; one which went beyond documenting women's experiences to think about the conceptual mileage in the term patriarchy; one which saw gender in terms of patriarchal gender relations; and which has its ultimate expression in the paper which I wrote with Jo Foord in 1986 (Foord and Gregson, 1986). It's important to stress, however, that this highly academic way of thinking and writing always co-existed uneasily in my mind with a feminism grounded in everyday life and day to day experiences, and with a scepticism borne of growing up in relative poverty in southeast London. For me, therefore, there was always a nagging inner voice which would never quite 'shut up'. Yes, I could see the advantages (and the flaws) of thinking in this grand theoretical tradition; yes, I could see the need for a feminist politics grounded in women's experience; but I could never quite get away from the fact that this was a middle-class (and white) political agenda, as remote from the everyday concerns of my mum and gran, and the world in which I grew up, as the workings of the Stock Exchange. If anything, as positionality has come to occupy a more central position in academic debate in the 1990s, these nagging doubts have assumed more and more significance in my thinking about 'gender'. Somehow I can never quite get away from feeling, to coin Janet Wolff's phrase, a 'resident alien' (Wolff, 1995); someone whose occupation labels them middle-class but whose working-class background is central to their identity, someone whose positionality continually has a nasty habit of triggering feelings of being forever 'out of place'. And increasingly, too, such ways of thinking shape the ways in which I think about 'gender'. In my academic writing and thinking, 'gender' is no longer a homogeneous category but one which in a recent project on domestic labour I have shown to be complexly and simultaneously interwoven with thoughts about class (Gregson and Lowe, 1994). In my current thinking, then, gender is not the significant difference; it's not even at the apex of social differences. Instead it's just one difference amongst many. But, at the same time, I continue to see the need for a strategic invocation of the importance of gender. Politically, and in teaching, this position continues to matter. Indeed, if it doesn't continue to do so, I feel that gender issues will cease to attract even the tokenistic gestures of recognition which they grudgingly gain in certain geographical circles.

References

Foord, J. and Gregson, N. 1986. Patriarchy: towards a reconceptualisation. *Antipode*, 18(2), 186–211.
Gregson, N. and Lowe, M. 1994. *Servicing the Middle Classes: Class, Gender and Waged Domestic Labour in Contemporary Britain*. London: Routledge.
Wolff, J. 1995. *Resident Alien*. Cambridge: Polity.

We begin our narrative of how feminist geographers think about gender with some of the first ways in which feminist geographers began to work with the concept.

Towards a geography of women... or gender is about women

One of the ideas which was introduced in Chapter One is that the content, assumptions and knowledge which passes as Geography are all reflective of the construction of the discipline by men. A good illustration of what is meant by this is the language deployed within geographical writing (see Section 2.4). Indeed, feminists working within geography have argued that although the language and the concepts in common use within geography may seem to be gender neutral, these concepts, and indeed the whole way in which we are encouraged to 'think geography', are grounded in masculine experiences and masculine realities. Furthermore, feminists have maintained that these masculine experiences are generalised and universalised; they are presented as the experiences of all. One of the main consequences of this has been to render invisible the experiences of women (and, we might add, everyone who is not encompassed by the categories white, male, heterosexual, middle class, able bodied). We can see the force of these arguments if we spend a few moments thinking about a few examples.

The first of these examples is 'work'. When we speak about 'work' and 'workers', for instance, in geography, all too frequently the implicit assumption is that we are talking about men and men's work. Our definitions and classifications of work, for example, all highlight and prioritise work performed by men. This is exemplified by the cartoon in Figure 3.5, and by the various classifications of occupations which are used to determine social class (you may have used these in analysing questionnaire survey returns, for example). All have at their heart a classification of occupations which both prioritises, and makes fine-grained distinctions between, the types of jobs which are (or increasingly in the case of manufacturing industry, were) performed for the most part by men in advanced capitalist societies. At the top of this hierarchy of occupations, then, are the male-dominated professions; whilst in terms of

Figure 3.5 Classifying occupations
Source The Oxfam Gender Training Manual

manual labour it is the type of work associated with the traditional male-dominated manufacturing industries (for instance, mining and engineering) which is highlighted and labelled as 'skilled'. In contrast, the myriad of jobs performed by many women (particularly those located in the service sector) are lumped together in a few categories, and are usually classified in the lower social class categories, as 'semi-skilled', or 'unskilled'. Diane Elson, writing in a development context, also emphasises the way in which apparently gender-neutral terms and concepts obscure a male bias. She argues:

> Rather than talking about women and men, and sons and daughters, use is made of abstract concepts like the economy, the formal sector, the informal sector, the labour force, the household. Or the argument is conducted in terms of socio-economic categories which on the face of it include both women and men, such as 'farmer' and 'worker'. It is only on closer analysis that it becomes apparent that these supposedly neutral terms are in fact imbued with male bias, presenting a view of the world that both obscures and legitimates ill-founded gender asymmetry, in which to be male is normal but to be female is deviant (Elson, 1995: 9).

In addition, most definitions and classifications of work, and most geographical analyses, equate work with paid employment. Such thinking not only tends to discount and/or trivialise women's paid employment (a phenomenon of massive importance in most contemporary societies) but ignores too the gender division of labour within societies (in which it is women who are primarily responsible for domestic and reproductive activities within and for the household). A consequence of this is that the contribution which unpaid forms of work make to economic productivity goes largely unacknowledged.

ACTIVITY

To reinforce these points about the implicit male bias imbued in the way geographers think about work, go through the following exercise. The Venezuelan data in Tables 3.1 and 3.2 illustrate the problems of learning about women's work from national censuses. Table 3.1 lists categories of work identified in a 1982 field study of 105 households on the Caribbean island of Margarita, a tourist resort and free port that is part of Venezuela. Table 3.2 is an English translation of the occupational categories used in the 1970 Venezuelan census.

- Try to assign each of the women represented by the information in Table 3.1 to one of the categories provided in the census (Table 3.2).
- How useful is the census list of categories for describing women's work?

A second example which is frequently used to demonstrate the masculine orientation of Geography is political geography. Here the state, government and institutions are located firmly within the public sphere. Furthermore, gender inequalities in society (expressed in men's and women's unequal access to and control over resources and decision making) mean that these same institutions tend to be male dominated and articulate male needs and interests. Nevertheless, such bias goes largely unrecognised and unacknowledged:

Table 3.1 Village women's work, Margarita Island, Venezuela (1982)

Housework
Cleaner in government offices
Weaves hammocks/ seamstress (repairs clothes)
Housework/ sells rabbits
Shoemaker in home
Housework/ makes and sells corn bread
Chambermaid
Housework/ crochets portions of hammocks for small manufacturer
Housework/ sells soft drinks from home
Housework/ weaves hammocks
Housework/ seamstress (makes clothes)
Laundress (operates from home)
Teacher
Street drink stand operator
Housework/ sells corn
School cook
Housework/ operates small general store in home
Housework/ rents space in home for small general store
Raises chickens and sells direct to consumer
Housework/ operates small fruit and vegetable store
Sells clothing on street in nearby town
Housework/ works in small family store/ makes parts of shoes
Sells clothing in store
Clothing store operator
Maid
Revendedore: retails clothing and housewares (purchased duty free) in streets and to
    private customers on the island and the mainland
Local government official
Housework/ takes in male boarder
Housework/ baby sitting

Table 3.2 Occupational classifications in the 1970 Census of Venezuela

Professional/ technical workers
Agents, administrators, directors
Office employees and kindred workers
Agricultural, livestock, fisheries, hunting, forestry, etc., workers
Transport and communication workers
Artisans and factory workers
Service workers
Others, not identifiable
Unemployed (identified according to categories above)

decisions made within the political sphere are presented as gender neutral and
as in the interests of (and taken on behalf of) a homogeneous 'public'. At the
same time, when women do make an appearance within political geography

this tends to be within the sphere of community politics. Not only is this form of politics frequently presented as less important, of lesser significance, than politics as expressed through the institutions of the state and government, but such actions are often characterised as in the 'private sphere' (i.e. they are about things like households, housing provision and form, and the quality of everyday life). Implicit within political geography, then, is a narrow definition of politics and political activity; one which thinks in terms of a hierarchy, in which male-dominated institutions, male needs and male interests are prioritised over and above those of women, who remain largely invisible as active political agents.

ACTIVITY

You should now turn to Readings A and B. Both of these provide further illustration of the arguments being made in this section. The questions posed below give you an opportunity to check not only that you have understood the specific arguments being made by Susan Hanson and Geraldine Pratt, and by Sue Brownhill and Susan Halford respectively, but provide you with the scope to use these papers as illustrative of the more general argument about the masculine content and assumptions of the discipline of human geography.

Reading A:

- How do Hanson and Pratt summarise the traditional portrayal of home and work in urban geography?
- With whom do they identify this traditional conceptualisation?
- What examples do they give of the elevation of work over home in urban geography?
- How do they argue that the links between home and work might be reconceptualised?
- How convincing do you find their arguments?

Reading B:

- How do Brownhill and Halford characterise traditional definitions of 'politics' and what has this meant for women?
- Notwithstanding shifting definitions of what is political, how has 'politics' been redefined so as to ghettoise women?
- What problems do Brownhill and Halford identify with the formal–informal dichotomy?

Recognising the implicit masculine bias prevalent within the various systematic divisions of Geography is vital to understanding how many feminist geographers began to work with the concept of 'gender'. Given the invisibility of women within geographical analysis and writing, feminist geographers set out to produce a counter to this, a geography of women (rather than of men) and one which emphasised just how different women's experiences are from male-defined norms. As a consequence, feminist geographers frequently came to equate the concept of gender with women (and women's experiences at that), whilst feminist geography itself came to be seen as fundamentally about women.

One of the most graphic examples of feminist geography as a geography of women – possibly *the* most graphic – is Joni Seager and Ann Olson's *Women in the World: An International Atlas* (1986). In this they argue that women's lives are strikingly different from those of men, and that their understanding requires a focus on women, rather than one which looks first at the world of men and then proceeds to women as an afterthought. Their project in this volume, therefore, is to map the world of women, and the result is a set of strikingly informative representations of a host of aspects of women's lives, including marriage, motherhood, work, resources, welfare, authority and body politics.

ACTIVITY

Figures 3.6 and 3.7 show the world distribution of 'battered women's shelters' (reflecting domestic violence) and the proportion of all paid workers who are women.

Looking at the map of women's shelters (Figure 3.6), try to answer the following questions:

- How would you summarise the pattern of provision of women's shelters (in which region(s) is provision apparently greatest/least)?
- Can you think of any potential reasons for this variation?
- What else strikes you about the level of provision and about data availability?
- Can you think of any reasons why data are classified as 'unknown' for so many countries?

Looking at the map of the paid labour force (Figure 3.7),

- In which areas do women comprise a high/low percentage of all paid workers?
- What sorts of reasons can you think of to account for these different participation levels?
- How accurate a picture do you think this is of women's work?
- Can you suggest any reasons why this map might be unreliable?

A second example of feminist geography as a geography of women is the Women and Geography Study Group's *Geography and Gender* (1984), the first introductory text in the field of feminist geography and, as such, a 'landmark' publication (see Section 1.1). This short text sets out from the premise that women have been hidden from Geography and seeks to show that there is a geography of women as well as of men. Much of the book, therefore, is an explicit attempt 'to redress the balance of introductory geographical material in favour of women' (p. 20), and the bulk of the book is taken up with chapters which look at women in the urban spatial structure, women's employment, women's access to facilities and women and development.

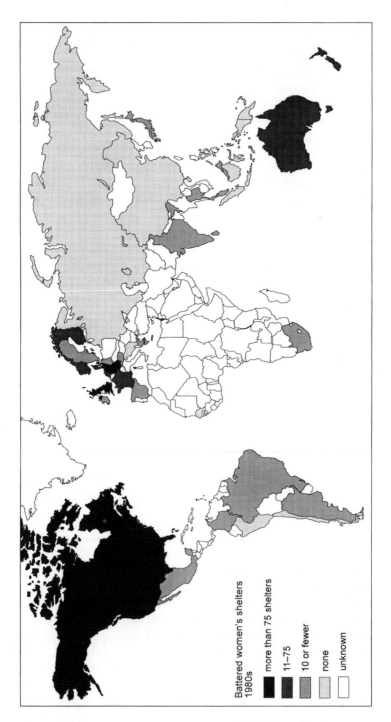

Figure 3.6   Domestic violence: battered women's shelters worldwide
Source Women in the World: An International Atlas

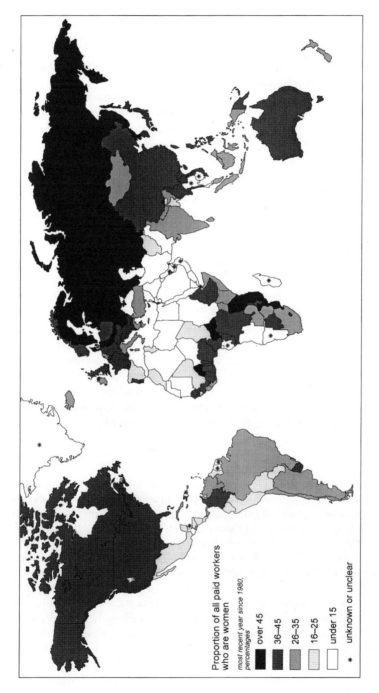

Figure 3.7   Labour force participation

Source Women in the World: An International Atlas

Read the Introduction of *Geography and Gender*, as well as *one* of Chapters Three, Four, Five and Six. If you decide to read Chapter Four, think about how this connects with some of the questions which we raised in relation to the labour force map from the *Women in the World* atlas.

Although the main emphasis in *Geography and Gender* is on gender as about women, and feminist geography as about producing a geography of women, the collective which produced this text did not see gender exclusively in these terms:

> Making women visible is simply not enough. The very fact that differences between women's and men's spatial behaviour patterns can be readily identified does not in itself guarantee that the geographer will do anything other than conclude: 'Well men and women are different, and it's interesting to see how this comes out in their behaviour'! In common with other approaches in geography which are critical of mainstream work we must analyse and understand why women remain in a subordinate position. (p. 20)

Such arguments show how, even at this earliest stage in the development of feminist geography, different ways of working with 'gender' existed, and in this case they co-existed within the same text.

In many respects the feminist geography which saw its task as about producing a geography of women can be seen as a strategic intervention in geography: it identified the male bias pervading geographical knowledge, labelled this as such and set about challenging this by producing geographical analyses which centred women, rather than marginalising them. It thereby made women visible for the first time within Geography.

ACTIVITY

We'd now like you to look at a selection of current first-year human geography textbooks. Look at the content of these and ask yourselves a few fundamental questions. Who is excluded and who is marginalised? What are the chapter headings? What are the key concepts within some of these chapters? And what are the assumptions which underpin these concepts? (It will help here to think back to some of the comments earlier in this section about work, workers and political geography.) What language is being deployed? Although this might at first sight appear to be gender neutral, is it?

Even a cursory examination of these first-year texts, we feel, should enable you to see both the way in which women continue to be largely invisible, and thereby marginalised, within contemporary geographical discourse, and the continued masculine bias within geographical writing. This situation is one which some feminist geographers see as highlighting a continued political need to produce a geography of women and, indeed, accounts for the ongoing commitment which many feminist geographers have to this earliest form of feminist geography.

There are, however, other feminist geographers who either disagree with this position or who, whilst agreeing with its strategic necessity, see some fundamental flaws in the project of constructing geographies based on women's experiences. These feminists argue that in focusing on women and their experiences, contrasting these with male norms, and in seeing women's lives as fundamentally different from those of men, such individuals rely on an essentialist view of women (see Chapter Two, Box 2.2, for a definition of essentialism).

Feminist geographers' diverse responses to the earliest feminist geography project of producing geographies based on women's experiences lie behind the emergence of a variety of different trajectories within feminist geography since the early 1980s. Thus, whilst many feminist geographers remained (and remain) convinced by the necessity and importance (politically and/or academically) of the geography of women project, others were not, and set about constructing alternative versions of feminist geography. As we now show, these versions still centre on the concept of gender, but they work with this concept in very different ways to those feminists researching the geography of women.

SUMMARY

- Gender has been used to refer to women; feminist geography can therefore be about producing geographies based on women's lives and experiences.
- This way of thinking gender reflects a desire on the part of feminist geographers to challenge Geography's traditional focus on the lives of men.

Thinking again about gender... gender roles and gender relations

Within the earliest forms of feminist geography, 'gender' – as we have seen – came to be equated with women, and specifically with describing and accounting for women's experiences. Although this position proved highly successful at describing women's experiences, accounting for them proved far more problematic within this framework. Thus feminist geographers moved quickly to invoke various concepts which appeared to give them a better explanatory grip on women's lives. It is this endeavour which lies behind the introduction of *gender roles* and *gender relations* to the feminist geography literature (see Boxes 3.1 and 3.2). Importantly, too, as well as seeming to offer more in the way of explanation, the introduction of these concepts is frequently portrayed as involving a shift away from the position which saw working with gender as exclusively about women. Indeed, all those working with these concepts espouse an interpretation of gender as involving *both* men and women.

Box 3.1   Gender roles

The concept of gender roles, grounded in recognised biological differences between men and women, explains differences between men's and women's lives in terms of socially constituted notions of what are appropriate activities for men and women. Some of the most widely used invocations of gender roles in everyday language refer to modes of behaviour labelled 'masculine' and 'feminine', for example strong/weak, assertive/compliant. However, in feminist geography the concept of gender roles is tied to men's and women's contribution to society. In Britain, therefore, it has become commonplace to label men's gender role as that of 'breadwinner', with women being 'homemakers' and childcarers. Alternatively, men are seen as 'breadwinners' with women and children as their 'dependants'. Feminist geographers, however, have emphasised how the development of these gender roles through the nineteenth century was embedded within an emergent spatial division of labour, with the male breadwinner role being integral to notions of waged employment beyond the home and the female homemaker role being equally important to the development of ideas about the home as 'haven' and as private space. More recently, as women have become increasingly involved in paid employment outside the home, and as men's employment has become increasingly threatened by redundancy, reference has been made to women's *dual role* and to *role reversal* households.

Further reading

Rose, D. and Mackenzie, S. 1983. Industrial changes, the domestic economy and home life. In Anderson, J., Duncan, S. and Hudson, R. (eds), *Redundant Spaces: Industrial Decline in Cities and Regions*. London: Macmillan, pp. 61–83.
Wheelock, J. 1990. *Husbands at Home: the Domestic Economy in a Post Industrial Society*. London: Routledge.

Box 3.2   Gender relations and patriarchy

The use of the term gender relations within feminist geography is closely linked with that of gender roles. Indeed, the concept was introduced in order to overcome some of the problems with gender roles, notably

- their inability to explain why roles take the form that they do;
- their failure to consider that roles are contested (as well as accepted) and to account for this; and
- their inability to cope with the ways in which roles change over time and space.

Rather than construing gender in terms of socially ascribed roles, then, this concept sees gender as a relational term, involving power relations between men and women. Although still tied to notions of male and female biological difference, it is male dominance and the processes which underlie this which constitute the main focus. This has led to various debates over the concept *patriarchy*.

Patriarchy

Patriarchy is a term you will find used in different parts of this book, and by a variety of different authors elsewhere. It can seem a confusing term at first because its meaning has changed over time and no one meaning dominates today. Weber used the term patriarchy to refer to a system of government in which older men ruled societies through their position as head of household, which gave them power over both younger men and all women. With the development of second-wave feminism, however, feminists both inside and outside geography started working with the concept of patriarchy. This work took a multitude of forms. Some authors continued to use the term to refer to men's domination of other men and women, but most feminist geographers used the term to refer to men's domination of women.

In 1986 and 1987 *Antipode* carried a debate about the conceptualisation of patriarchy in feminist geography. Jo Foord and Nicky Gregson (1986) thought feminist geography lacked an overall theoretical framework and attempted to contribute to the development of such a framework by reconceptualising patriarchy within a realist framework (see Cloke *et al.*, 1989, Chapter 5 for a discussion of realism). They argued that gender relations are trans-historical and trans-spatial; that they exist at all times and places. They identified the existence of gendered men and women as the basic characteristics of gender relations, and biological reproduction and heterosexuality as the necessary relations. In other words, all you need to have gender relations is gendered men and women, and the key relations between these gendered men and women are biological reproduction and heterosexuality. However, they argued that just because gender relations are trans-historical and trans-spatial does not mean that they are the same in every time and place. These relations could, for example, work to the advantage of men, work to the advantage of women, or be egalitarian. Thus Foord and Gregson argued that patriarchy was one particular form of gender relations, a form in which men dominate women. They therefore called these patriarchal gender relations.

A flurry of articles followed this paper. Most authors had some points of agreement with Foord and Gregson, but differed with their analysis in a number of respects. Louise Johnson (1987) and Jaclyn Gier and John Walton (1987) questioned Foord and Gregson's desire to develop an overall theoretical framework for feminist geography and disputed the value of realism as a way of achieving this. Linda McDowell (1986) and Lawrence Knopp and Mickey Lauria (1987) challenged Foord and Gregson's analytical separation of gender relations and mode of production, of patriarchy and capitalism. They argued that the study of women's oppression could not be separated from a class analysis.

Bibliography

Cloke, P., Philo, C. and Sadler, D. 1989. *Approaches to Human Geography*. London: Paul Chapman.

Foord, J. and Gregson, N. 1986. Patriarchy: towards a reconceptualisation. *Antipode*, 18(2), 186–211.

Gier, J. and Walton, J. 1987. Some problems with reconceptualising patriarchy. *Antipode*, 19(1), 54–58.

Gregson, N. and Foord, J. 1987. Patriarchy: comments on critics. *Antipode*, 19(3), 371–375.

Johnson, L. 1987. (Un)realist perspectives: patriarchy and feminist challenges in geography. *Antipode*, 19(2), 210–215.

Knopp, L. and Lauria, M. 1987. Gender relations as a particular form of social relations. *Antipode*, 19(1), 48–53.

McDowell, L. 1986. Beyond patriarchy: a class based explanation of women's subordination. *Antipode*, 18(3), 311–321.

Walby, S. 1989. Theorising patriarchy. *Sociology*, 23(2): 213–234.

Walby, S. 1990. *Theorising Patriarchy*. Oxford: Basil Blackwell.

One of the best examples of how feminist geographers have utilised the concepts of gender roles and gender relations comes in Linda McDowell and Doreen Massey's article in *Geography Matters*, entitled 'A woman's place' (McDowell and Massey, 1984). Defining patriarchy in terms of male dominance, they argue in this chapter that different forms of economic development in different regions and/or local areas provided different challenges to male dominance. In their words:

> ...capitalism presented patriarchy with different challenges in different parts of the country... this process of accommodation between capitalism and patriarchy produced a different synthesis of the two in different places. It was a synthesis which was clearly visible in the nature of gender relations, and in the lives of women. (p. 128)

Using four case study examples – the Northeast coalfield, the Northwest, the Fens, and Hackney in inner London–McDowell and Massey show how gender relations and gender roles vary over space and how they are constituted in place. The nineteenth-century Northeast coalfield, for example, is cited as an extreme instance of patriarchal gender relations: an area dominated by highly distinctive male and female gender roles, with men selling their labour to private mine owners and women responsible for 'servicing' their menfolk – washing, feeding, clothing husbands, sons, fathers and brothers in preparation for the next shift. In contrast to the Northeast, the nineteenth-century cotton towns of Lancashire were places where women's gender role was far from exclusively confined to the domestic sphere. Here, as well as being responsible for the home, women were employed in the cotton factories as weavers, a situation which is considered to have played a major part in the development of feminist politics in the Northwest region. Turning their attention to the twentieth century, McDowell and Massey go on to show how nineteenth-century gender roles provide the basis for understanding something of the changing fortunes of the Northeast and the Northwest through the period since 1945. Northeast women – with a tradition of non-paid employment – represented a classic instance of 'green labour', and indeed, it is *their* labour – rather than that of Northeast men – which proved attractive to many multinational companies seeking a UK location in the

1960s and 1970s. In contrast, the militancy of the women from the cotton towns and their tradition of paid employment out of the home (together with the lack of regional aid granted to this area in the post-war period) is argued to be behind increasing levels of female redundancy in the Northwest since the 1960s. In McDowell and Massey's work then, place-specific gender roles grounded in particular historical periods are shown to provide the conditions for the emergence of modified gender roles in subsequent periods.

A second example, which places considerable emphasis on the concept of gender roles, is Jackie Tivers' book, *Women Attached* (1985). The focus of this book is on the daily activities of women with young children, and reflects Tivers' concern to challenge the continuing marginality of women's activities within geography (see previous section). The study is based on 400 interviews with women in the Borough of Merton in southwest London and illustrates the extent to which women with young children found it difficult to obtain paid work, even if they wanted to – largely because of poor childcare facilities or limited mobility. Those women who did work worked in low-paid jobs within close proximity to their homes. Daily activity patterns were structured by the constraints on mobility presented by the presence of young children and the lack of access to a car.

Although there is some variation in the lives and opportunities of the women interviewed according to social class, particularly in terms of personal mobility, the study emphasises the extent to which the ideology of gender roles influences the activity patterns described by the empirical data. Thus the employment possibilities, the kinds of unpaid activities undertaken, and the daily activities of the women with young children interviewed, are seen as defined within the 'societal constraints' imposed by the ideology of gender roles. In other words, Tivers sees gender differentiation, based on recognised biological difference, as the main constraint upon the activities and choices of the women interviewed: the 'gender role constraint' assumes that the primary job of childcare and homemaking will be undertaken by women.

Our third example of work which sought to use the concepts of gender roles and gender relations to account for women's experiences is Janet Momsen and Janet Townsend's edited volume *Geography of Gender in the Third World* (1987). This book looks at the significance of gender divisions within the 'Third World' and attempts to examine what the editors acknowledge to be an enormous variation between countries in the relationship between men and women and the environment. In addition, they argue that this focus is not simply to 'add women in' to development, but to recognise the extent to which the incorporation of gender issues challenges existing development theory and practice.

The book seeks to depict a regional geography of gender (p. 81), involving the recognition of both continuities and diversities. Gender here is defined as socially constituted, and consequently as varying between societies, whilst the social constitution of gender in different societies is defined through the concept of gender roles. The book's focus is on the working lives of women – as they are involved in productive and reproductive activities – and the ways in which their lives differ from those of men. These differences, as in the above examples, are interpreted primarily through the lens of gender roles. However, although emphasising diversity, the book also stresses commonality:

> Despite geographical variations, class variations and individual variations, the worldwide theme of the geography of gender is female subordination. The geography of gender has both this world-wide continuity and regional and local diversity... All over the world women's work tends to be defined as of less value than men's and women seem to have far less access to all forms of social, economic and political power. (p. 28)

Now, one of the points which we find particularly interesting about the ways in which feminist geographers have used the concepts of gender roles and gender relations is the way in which, whilst gender is no longer taken to refer exclusively to women, and whilst men are admitted to and indeed seen as central to analyses, for feminist geographers critical of the geography of women project, gender still tends to translate into being about *women's* roles, *women* in relation to patriarchy.

Part of the reason why feminist geographers working with gender roles and gender relations have in practice interpreted these as being about women is undeniably connected with feminist politics and reveals the connections which these ways of working have with the geography of women project. Here (as there) the emphasis is strategic: women's gender roles, their position within patriarchy, comes under the microscope (so to speak) in order that *women* can see, articulate, understand and hopefully challenge the conditions behind (and productive of) gender inequalities. However, one of the problems with this position academically is that such thinking tends to homogenise and essentialise the qualities and characteristics identified with men and women, male and female. Indeed, no matter how hard feminist geographers tried to nuance their accounts of gender roles and gender relations with qualifying remarks about historical and spatial specificity, the tendency to read accounts as applying to *all* women, and as characteristic of *all* women, was there. To counter this tendency feminist geographers started to consider *gender differences*, but within this have focused particularly on differences *between women*.

SUMMARY

Gender has been used to refer to understanding and explaining the differences and inequalities between men's and women's lives.

Gender and other social differences...or gender is still about women and men but also admits that there are differences between women and between men

Amongst the many other differences which feminist geographers have addressed in thinking about differences between women are those of life-course, race, class, sexuality and place. Importantly, though, although emphasising differences between women, feminist geographers have still prioritised gender as their central analytical category. Thus, within this position gender remains the *primary* social relation on which experiences are based and identities constructed. Or in more specific terms, a person is first a man or woman, a difference which determines his or her core identity, with race, ethnicity, class, sexuality, etc., being added on as contributing to this core identity. We maintain that what this establishes in feminist geographers' writings is a *hierarchy of social differences*, with gender at the apex.

The following represent a few examples of this position within feminist geography, in which feminist geographers have considered other differences alongside those of gender. However, it is important to stress at this point that, in identifying particular pieces with specific other differences, we are not saying that these pieces are exclusively concerned with gender and one other social difference. Indeed, if you delve further into them you'll find that manifestly they are not. Multiple differences appear within many, if not all, of these pieces, but what connects them all is their reliance on gender as the primary analytical category.

The importance of the lifecourse

Relatively unexamined in both the feminist and geographical literature, the importance of *lifecourse differences* forms the basis of Cindi Katz and Janice Monk's edited volume, *Full Circles* (1993). Prompted by their own work on children and on older women respectively, Katz and Monk aim in this collection to move analysis away from the traditional feminist concern with the lives and experiences of women in the childbearing/rearing years 'to describe and interpret the geographies of women's lives in an array of settings from the perspective of the lifecourse' (p. 4). Drawing on work conducted in North America, Australia, the Caribbean, Latin America and the European Union countries, the various essays contained within this collection provide abundant illustration of the diversity and commonalities of women's experiences across the lifecourse, and testify to the importance of context (of space and place) in shaping how women experience the lifecourse. What Katz and Monk (and their various contributors) are attempting to do in this volume, then, is to use the concept of the lifecourse to challenge the essentialism inherent in feminism's (and feminist geographers') traditional emphasis on women in the childrearing years and the consequent (inadvertent) portrayal of these years as *the* defining experience common to all women.

ACTIVITY

At this point it would be useful to look at the introduction to *Full Circles* and two of the chapters. The chapters span various contexts, so it is possible for you to use your own interests to select two for further reading. As you read these, ask yourself a few questions:

- How open to difference is the representation of the experience of the lifecourse in the contributions which you've read? Think here about whether the authors emphasise universality of experience within various lifecourse stages or whether they stress that other lines of social difference (for example, class, race and ethnicity) complicate the notion of the lifecourse.

● As you read these chapters, can you identify any problems with the concept of the lifecourse?

Although *Full Circles* makes an excellent case for the importance of the lifecourse to understanding the diversity of women's experiences and provides ample illustration of the importance of space and place in shaping such diversity, one of the points which strikes us is the continued primacy accorded to gender in this volume. Thus, despite the emphasis on the lifecourse, diversity and context, it is gender – and gender interpreted in terms of the geography of women's experiences – which sits as the heart of this text. This is a good example of the point which we made in the introduction to this section concerning the importance of seeing the development of feminist geographers' ways of working with gender in a non-linear way. At the same time as emphasising the importance of the diversity of experience within the category woman, *Full Circles* maintains a commitment to the 'gender equals women' position and sees the task of feminist geography as that of producing a geography of these diverse experiences.

The importance of place

Much like *Full Circles*, Janet Momsen and Viv Kinnaird's edited volume, *Different Places, Different Voices* (1993), provides another example whereby intra-gender differences are acknowledged, recognised and explored. However, unlike *Full Circles*, it is place, rather than the lifecourse, which is seen as critical to understanding and determining women's diverse experiences. Organising place according to global regions – Africa, Latin America, South Asia, etc. – the book uses various case studies to explore the diversity of women's lives and livelihoods, raising issues along the way of race, ethnicity, class and colonialism.

ACTIVITY

At this juncture we suggest that you read the introduction to *Different Places, Different Voices*, together with one chapter from each of the global regions sections. As you do this, ask yourself to what extent the chapters show heterogeneity and the explanatory importance of difference in space and time.

One of the things which we find particularly interesting about *Different Places, Different Voices* is the way in which, despite the emphasis accorded to place, it is 'gender' which remains the primary, central and unifying analytical category within this volume. Indeed, it is gender interpreted in terms of the diversity of women's experiences which is continually stressed. In many ways, then, the emphasis in this volume is much as in *Full Circles*: it provides further testimony to the continued commitment of many feminist geographers to the 'gender equals women' position and to feminist geography as a geography of women.

The importance of class and race
A good example of a piece of work which makes reference to the importance of class and race differences between women is Gill Valentine's (1989) paper, 'The geography of women's fear', based on research conducted in Reading, Berkshire (see Reading C). In this paper the author focuses on women's fear of male violence and their perception and use of public space. Various types of public space are highlighted as perceived by women to be 'dangerous places' at 'dangerous times', notably open spaces (for example, parks, woodland, waste ground, canals, rivers and countryside) and closed spaces (for instance, subways, multistorey car parks and alleyways). However, in the paper Valentine also makes reference to differences between the two groups of women she interviewed, one group of whom lived on a private, middle-class housing estate on the edge of Reading and the other on a working-class council estate. As she says here:

> In Reading both the middle and working class white women interviewed hold an image of a predominantly Afro-Caribbean residential area as dangerous for white women because of a racist assumption about the violent nature of black males. Similarly, the middle class women also anticipate a large 'rundown' council estate to be rough, whereas the residents of that area perceive themselves to be safer than the middle class women do in their own housing area. (p. 388)

Class and race, then, are evidently important in shaping how safe women feel in particular built environments. However, much as with the previous two examples, the overriding emphasis in Valentine's paper is on the commonality of women's experiences. All women are argued to be fearful of male violence, to associate this with certain types of public space and to inhibit their use of space accordingly. Gender then is *the* significant difference in this paper. Furthermore, although fear of male violence plays a central role in this piece, 'gender' itself is predominantly interpreted in terms of the geography of women.

ACTIVITY

You might like to organise a discussion amongst yourselves over this article. Specific issues which are likely to provoke debate, which you might like to think about and which you might disagree over are:

- The representation of men in this paper
- The lack of attention given to violence against men
- The way in which Valentine uses fear of male violence to account for women's need for a male partner
- Women's behaviour in public space. Is this any different from that of individual men's?

Now, rather than taking the gender and other social differences position at face value, it can be argued that, in placing gender at the apex of a hierarchy of social differences, feminist geographers have once more been working strategically: although they present gender as the most significant social difference, this does not necessarily mean that this is how they really see things.

Rather, they could be argued to be elevating gender for political reasons, to ensure that its importance is on the agenda, acknowledged and recognised. In many respects, we feel that this is precisely what many feminist geographers who work with gender in this way have been (and are) doing. In effect they are deploying a tactic known as strategic essentialism (see Box 3.3).

Box 3.3   Strategic essentialism

Although many feminists have resisted essentialised attributes assigned to women, arguing that women are not a homogeneous group and that greater attention needs to be paid to *difference*, others have chosen to emphasise the similarities between women's experiences. Choosing to emphasise women's common experiences and needs is a strategic choice because it unites women as one group working for social, political and economic advance. In this position, difference, although recognised, is suspended in favour of a strategic alliance seeking political, social and economic change.

However, increasingly this position is being challenged by other feminists, who argue that working with gender in a way which grafts other social categories onto the category gender is inherently flawed. For such writers, according primacy to gender as the key analytical category means that gender is still being seen as about similarities, and specifically as about the way in which women (despite their other differences) are united by being gendered as women. So what alternative way of working with gender do this group of writers offer?

SUMMARY

Some feminist geographers argue that in working with gender we need to be sensitive to differences between women. However, for some feminist geographers, gender is still the most significant difference.

Decentring and destabilising gender

One position which has recently emerged within feminist geography is largely informed by postcolonial feminist theories. Postcolonial feminist theory is a large and complex body of work which explores the interrelationships between identity, knowledge and power. It is described as postcolonial because the particular historical and geographical context with which these theorists engage is that of the colonisation of what we now call the Third World from the sixteenth century onwards. These theorists argue that this is the crucial context for understanding the construction of identities and difference even now, and their critical goal is to move towards understandings of identity and difference which are not structured by that context: understandings which are postcolonial. Several strands of these arguments are particularly relevant to feminist geographers but here just two will be teased out. These are, firstly, the ways in which postcolonial feminist theory offers a critique of the kind of knowledge constructed by Geography about other places; and secondly, how postcolonial feminist theory challenges feminist geographers' understandings of gender.

Postcolonial feminist theories are especially concerned with what kind of knowledges about colonised people and places were constructed by the colonisers. Geography as a discipline was active in this process of knowledge construction; meetings of local geographical societies were very popular in the nineteenth and early twentieth centuries, attracting large audiences for talks illustrated with lantern slides about various aspects of colonial life. Postcolonial writers argue that in these sorts of fora, the colonised were shown as the Other of the colonisers. In these lecture halls and town halls, the superiority of home was being established by showing the inferiority of the exotic. But the kind of knowledge being produced about such places presented itself as rational, scientific, objective and neutral. Explorers and travel writers, geographers and anthropologists, represented themselves as observers merely describing what they saw. They adopted the distanced tone of voice and writing which the previous chapter described as masculine, but which postcolonial writers argue in this context must also be seen to be part of a certain kind of *white* masculinity. White men produced knowledge about these places, and their assumption was that only they could do this. Only rational, white men were able to produce reliable, scientific evidence about the world. Other sorts of people – women, 'natives' – could not. They did not have the requisite powers of reason or observation; nor were they strong enough to endure the trials and tribulations that made the white masculine traveller into an adventurous, heroic explorer. Thus their apparently simply descriptive mode of knowledge obscures the claim to authority being made by these observers. Their knowledge was constructed not as *a* knowledge but as *the* knowledge; they produced what they imagined to be comprehensive, totalising forms of knowledge which often characterised those colonised in essentialist terms. It is these totalising, essentialising and Eurocentric forms of knowledge which postcolonial writers are challenging.

These arguments from postcolonial feminist theories have also been used by some feminist geographers in order to understand the ways in which Geography as an academic discipline might be described as masculinist. As well as only looking at men's experiences and men's interests and generalising from them (as described earlier in this chapter), some feminist geographers have argued that there is something about the very structure of Geography's knowledge that continues to be shaped by this colonial legacy. In its often highly authoritative claims to present the world simply as it is, in its traditions of heroic fieldwork, in its continued reluctance to admit the voices of its Others, Geography remains a colonial enterprise.

This kind of critique is also the backdrop for the work of those feminist geographers concerned to recover the lives of women who were producing geographical knowledges during the colonial period. But here we encounter the second aspect of postcolonial feminist theory most relevant to feminist geographers. This is an aspect which may already have suggested itself to you, since it was broached in the previous chapter. In Chapter Two, you were asked to think about extracts from two accounts of the travels of Mary Kingsley, a middle-class English woman who travelled to Africa in the 1890s. While one extract suggested that Kingsley's gendered position meant that she could not appropriate what she saw into an authoritative and totalising view because that kind of view was a masculine one, the other extract argued that, as a white woman, Kingsley was, and indeed could only be, complicit with the white colonial appropriation of Africa. This raises in very direct form the theoretical question posed to feminist geographers by postcolonial feminist theory: is gender always the most important analytical category? Or are there moments and places where other kinds of social difference become more important? In relation to the example of Kingsley, is it more important to consider her gendered or her racialised position when thinking about the geographies she made? Not only does this theoretical question challenge the positioning of gender as the primary social relation and central analytical category within feminist geography, but it also questions the ways in which feminist geographers have thought about gender. As such, this position is best described as one which decentres and destabilises the analytical category gender.

The critique which this position offers of ways of working with gender is particularly targeted at the 'gender and other social differences' position (see previous section), and is three-pronged. First, those influenced by these arguments maintain that seeing gender as the most important category of social differentiation, and grafting on to this other social differences (as discussed above), effectively homogenises and simplifies the complex and diverse experiences of men and women. Seeing individuals as primarily male or female is argued to conceptualise other social differences as supplementary and marginal to the primary gendered distinction. As a consequence, it is argued that there exists no scope within this way of thinking to understand how other social relations may shape, inform or even transform the gendered nature of our experiences and identities.

A second problem identified with working with gender as the most significant social difference is the assumption that it is possible to identify and separate out different and distinct components that make up an individual's identity, and then identify which experiences emerge out of gender differences and which are shaped by, for example, race and sexuality.

Finally, there is the problem of the normalising of particular identities which goes on with the 'gender and other social differences' position: whilst individuals are differentiated by gender, they are at the same time assumed to be white, heterosexual and middle-class. This can be seen once we recognise that it is only those who do not fit into these normative characteristics for whom 'other differences' become relevant or appropriate. Instead of adding race, sexuality and class to the category gender, we are actually only required to add those subcategories which are not the norm, that is (Black) race, (homo) sexuality and (working) class. Implicit here, then, is the assumption that you do not have to be explicit about the ways in which these other differences impact upon your experiences as a man or woman if you are white, heterosexual and middle-class.

The above three problems, namely according primacy to the category gender in feminist analysis, arguing that it is possible to separate out the influence of different social categories and the normalising of particular identities, are encapsulated in the following quote from Dionne Brand:

> I remember a White woman asking me how you decided which to be – Black or woman – and when. As if she didn't have to decide which to be, White or woman, and when. As if there was a moment that she wasn't White. She asks me this because she sees only my skin, my race and not my sex. She asks me this because she sees her sex and takes her race as normal. (Brand, 1990: 46, quoted in Jackson and Penrose, 1993)

The argument here is that not only is it difficult to distinguish between the effects of different social relations, as we have seen in the above quote, but it is impossible to talk about gender without considering how gender itself is constituted through other social differences. One of the clearest expressions of this way of thinking occurs in Vron Ware's *Beyond the Pale*, where she states:

> ...to be white and female is to occupy a social category that is inescapably racialised as well as gender. (Ware, 1992: xii)

Gender, then, is implicitly racialised: that is, it is constituted racially. Arguing along similar lines, Sarah Radcliffe (1994) suggests that feminist geography often writes about Third World women as an undifferentiated mass rather than as diverse and active subjects. She argues that there is a need to develop a framework for analysis which could focus on how different masculinities and femininities are created and reproduced over time and space (Radcliffe, 1994). Thus, rather than taking gender as a natural category, feminist geographers have begun to rethink the category of gender. In other words, they have begun to present the idea that there is no one masculinity or femininity; instead there are masculinities and femininities. This line of analysis emphasises that race, gender, class and sexuality are mutually constituted.

If the arguments made by postcolonial feminist writers offer a number of challenges to feminist geographers, so too do post-structuralist feminist arguments which focus on the process of Othering. As we have already noted, feminist postcolonial writers argue that during the colonial era, the colonised were positioned as the Other of the colonisers. However, these critics also argue that many other kinds of difference were interpreted by the colonisers through this structure of an Other. Black was contrasted to White, barbaric to civilised, violent to moral, colony to home, feminine to masculine, proletarian to bourgeois, rural to urban. In each of these cases, the Other was made sense of only as the opposite of the powerful. The imperial powers attempted to refuse the Other a voice of its own, tried to deny it a language with which to define its own position. Since this process of constructing knowledge was so important to colonialism, many postcolonial critics attempt to challenge it. They try to shift ways of making meaning away from this dualistic structure of the powerful and their Other (see Box 3.4 for a discussion of dualisms).

The ability of the powerful to colonise meaning should not be exaggerated, and, broadly speaking, two critical strategies have been deployed against it. The first is to search for knowledges not entirely colonised by dominant ways of knowing. Part of the project of the black feminist writer bell hooks, for example, is to find a voice to express her sense of identity and place. This voice is a complex one. It speaks as at once black, female and working-class, a voice which can 'vividly recall efforts to silence my coming to voice' (hooks, 1990, 147); it is a voice whose silences acknowledge the effects of being marginalised by both racism and patriarchy; and it is a voice which moves between different registers and different media. Speaking of her efforts and those of others to speak different knowledges, she says:

> It is no easy task to find ways to include our multiple voices within the various texts that we create – in film, poetry, feminist theory. Those are sounds and images that mainstream consumers find difficult to understand. Sounds and scenes which cannot be appropriated are often that sign everyone questions, wants to erase, to 'wipe out'. I feel it even now, writing this piece when I gave it talking and reading, talking spontaneously, using familiar academic speech now and then, 'talking the talk' – using black vernacular speech, the intimate sounds and gestures I normally save for family and loved ones. Private speech in public discourse, intimate intervention, making another text, a space that enables me to recover all that I am in language, I find so many gaps, absences in this written text. To cite them is at least to let the reader know something has been missed. (hooks, 1990: 147)

hooks is suggesting that language itself can be a site of struggle: who speaks, how and where are issues structured by power relations. These issues were raised in Section 2.4 in Chapter Two: feminist geographers have also had to think about how to write and speak in different contexts in order to be heard. hooks goes on to argue that a new voice must be found which structures its understanding of difference not through the notion of an Other, but differently, more respectfully and less oppressively. Sometimes she advocates strategic essentialism as a way of articulating a voice that challenges the power- ful; but her larger project is about finding a more complex sense of identity which can negotiate difference from itself without pathologising it as an Other.

A second tactic adopted by many postcolonial feminist writers, like some other feminists, is to turn to post-structuralism as part of their efforts to challenge this process of making meaning through contrasts. Post-structuralism is a way of interpreting knowledge which tries to displace the contrasts through which Western knowledges in particular tend to interpret the world. Feminist writers inspired by post-structural, postcolonial arguments suggest that the structure of Othering itself can be unstable. To pick up again the arguments made a few paragraphs previously, the different dimensions of social position structured through Othering – race, gender, class, place – clearly mediate each other. As we have also seen, this produces some very difficult analytical questions. These Others in the colonial (and contemporary) context do not neatly align; as the difference between Blunt's and Pratt's interpretations of the Victorian woman traveller Mary Kingsley showed (see Section 2.3 in Chapter Two), a white woman in Africa is 'Othered' in some ways but not in others, and this produces disagreement over the consequent complexities when trying to locate her social position. These complexities can be the focus of critiques attempting to challenge the coherence of knowledges claiming universal knowledge about the world, however. Instead of acknowledging the power of dominant know-ledges, feminists and other critics can explore its weaknesses, its failures, its contradictions, its instabilities.

An example of a feminist geographer attempting such a critique of domi-nant geographical knowledges can be found in an essay by Jane Jacobs (1994; see also G. Rose, 1993), which examines the complex intersection of colonial-ism, patriarchy, feminisms and environmentalisms in contemporary Australia. Jacobs is writing about the struggle which took place through the 1980s over whether an Aboriginal sacred site in central Australia – a sacred site for Arrernte women – should be flooded as part of a recreational lake and flood mitigation dam scheme. As Jacobs shows, this struggle produced some extremely complex political alignments. Some of these alignments involved the racialised Othering of Aboriginal people as, for example, much closer to the Earth than their white feminist and/or environmentalist supporters; at other moments, Aboriginal women in particular were singled out and Othered as much closer to nature (Chapter 7 returns to this issue). However, Jacobs refuses to read this struggle 'only in terms of the reiteration of a politics of Western, masculinist supremacy' (1994: 169). She argues that to describe this struggle only as the effects of various processes of Othering both overestimates the stability of those Othering processes, and underestimates the active politi-cal interventions undertaken by Aboriginal men and women. Instead, she examines what she describes as 'a political interspace' in which the distinctions between Aboriginal and non-Aboriginal are complex, confused and unstable. She points out that some Aboriginal women were happy to ally with white feminist environmentalists for the sake of saving the sites while others were less so, for example. Her conclusion, however, is that although such coalitions between and among Aboriginals, non-Aboriginals, environmentalists and femi-nists did eventually save the sites from flooding, they 'nonetheless resonate[d] with less sympathetic moments in the history of settler Australia' too (Jacobs, 1994: 191). Jacobs' discussion of this 'interspace', then, like hooks' efforts to find a different voice, tries at once to move beyond dominant modes of Other-ing while also acknowledging the ways that those modes continue to make their mark on contemporary struggles. In the process, gender as a central and a stable analytical category becomes displaced.

Interesting and exciting as this type of writing is, there is little doubt that its appearance within feminist geography has triggered a considerable hiatus of debate. At the heart of this debate is the question of how feminist geogra-phers respond to the central problem thrown up by the destabilising gender position. The problem is this: *arguing that gender is mutually constituted with race, sexuality, class and so on not only challenges the primacy of the concept gender but destabilises how we think about gender too. If we cannot separate out the parts of our identity which are based on gender from those associated with race, for example, then what does this mean for the project of feminism and for the various feminist geography projects which have taken gender (albeit variously interpreted) as their central analytical category?* The implications of this position will be discussed more fully in the conclusion to this volume.

SUMMARY

- Some feminists argue that working with gender as the most significant difference homogenises and simplifies the diversity and complexities of individuals' experiences.
- Some feminists argue that using gender as the primary social and analytical category makes certain assumptions about the ways in which gender, race, class, sexuality and so on shape people's lives.
- Some feminists argue that social categories are not separate and distinct but are mutually constituted.

3.4   Summary discussion

In the previous section we showed that there are four main ways of working with gender in feminist geography, but that these have not developed in a chronological fashion. Instead they co-exist, within feminist geography at any one time, within the writings of individual feminist geographers and even within the same texts. Inevitably then, there is no easy answer to the question 'how have feminist geographers worked with gender?', and there are many differences between feminist geographers themselves. Indeed, we who have produced this chapter do not share the same ideas about how to work with gender, although we do agree with how the different positions have been represented here. At this point, therefore, we suggest that it would be useful to spend some time thinking about these positions, the differences between them, and which way of working with gender appeals most to you. The following activity has been designed to help you in this.

ACTIVITY

As a class, think of a research topic for further investigation. Utilising the four ways of working with gender examined in Section 3.3, work out the different ways in which you could approach this topic and evaluate the differences between them. In making this evaluation it might be useful to think about how successful these representations are of the topic you chose to investigate; what and whom these representations include and exclude, and how they tackle difference. Does your evaluation of these positions depend on who you are?

READING A

Hanson, S. and Pratt, G. 1988. Reconceptualising the links between home and work in urban geography. *Economic Geography,* **64(4), 299–318.**

The relationship between home and work – residential location and work location or housing markets and labor markets – has been central to urban geography and to the models that both reflect and delineate our vision of the city. Although the critical importance of the home–work link has long been well recognised, we argue here that the way in which this central relationship has been conceptualised has changed relatively little over the years, despite several weaknesses in the prevailing conceptualisation... We argue that within human geography home and work have been viewed as separate spheres and that the division of academic labor within urban geography itself has nurtured and sustained this schism... Little work in urban geography has spanned these two arenas... It is now commonplace to point out that home and work have been viewed not only as separate but also as gendered, with the workplace epitomising the male realm and the residence, the female... [Furthermore] the ways in which home and work are seen to be related and the ways in which each sphere affects the other have been modelled separately for women and for men. Characteristics of women's labor force participation are believed to be filtered through their home experiences while the opposite tends to be the case for men... Also, women's job satisfaction...has been seen as a function of their family roles and the problems of coping with the double burden of home and work, whereas men's job satisfaction has been seen as being related explicitly to job conditions. On the other hand, the home is often conceived as a 'haven' for male workers, as a retreat from and solace for degraded work conditions. The meaning of the home for men is, therefore, conditioned by their work experiences...

Our point is not that the home–work link has been overlooked, but that it has been conceptualised in a limited and limiting way that reflects a fundamental, underlying view that the two spheres are essentially separate... (Here we) identify the particular aspects of that legacy that are in need of rethinking namely: (1) the conceptualisation of both 'home' and 'work' simply as points in space and the linkage as a line joining these points; (2) the priority given to the workplace in the establishment of the home–work link; (3) the idea that all households are identical and use the same criteria in making the residential and workplace location decisions that define the home–work linkage; and (4) the notion that variations in local context do not essentially affect the nature of the work–home tie...

In our view, the priority given to the work location...is one example of the gender bias inherent in (urban geography) models. A more fundamental gender bias exists in what is taken to constitute work, namely paid employment... Gender, as well as class, bias is also apparent in these modellers' implicit assumption that the urban population is made up of essentially one household type, consisting of one (male) worker and one (female) full time home-maker... There is little doubt...that urban modellers overgeneralised the white middle class experience... [Their] models have served to rigidify our way of thinking about cities, to reify the gender division of labor, and to reinforce the status quo...

In the traditional models of uban spatial structure both labor force partic-ipation and the location of work are simply assumed as given. In our view, these assumptions reflect the male origins and orientation of these models... The home environment can be important to the work decision in at least three ways: (1) as a source of potential employment opportunities; (2) as a source of support services, and (3) as an agent of socialisation... For many ...women, the residential location is considered fixed; the job search proceeds not from a residential location that can be moved to accommodate a new job site but from a given home site from which suitable employment must be found... The home environment can affect work decisions...insofar as it functions success-fully or unsuccessfully as a social resource. A number of social supports and services – such as day care for children and the elderly; transportation; coun-seling, retail, service and recreation facilities – may be present or absent to varying degrees in the local neighbourhood, and their presence or absence can affect a person's decision of whether or not to work and where... Beyond the local availability of jobs and formal and informal services, a third way in which the home environment can affect work is through the socialisation process by which certain work related attitudes, skills and goals are passed from one generation to the next. The process occurs not only within the home but also within the neighborhood via schools and social interaction...

In our view, the concept of home – as it affects work – should be a good deal more than just a point in urban space. A reconceptualised 'home' would be expanded outward to include the surrounding neighborhood and inward to include intra-household interactions. (This) would recognise the neighbor-hood as the locus of a set of potential jobs, social networks and services that bear critically upon the household's work decisions... Expanding the concept of home inwards would open up the household itself to examination of the ways in which intra-household negotiations, interactions and strategies affect the work decisions of its separate members...

READING B

Brownhill, S. and Halford, S. 1990. Understanding women's involve-ment in local politics: How useful is a formal/informal dichotomy? *Political Geography Quarterly*, 9(4), 396–414.

Historically the term 'politics' has been used to describe little more than mainstream party politics taking place within the institutions of the state or, sometimes, at the workplace. This definition of politics as public power, and in particular as power emanating from the state, has some major implications for women. Feminist analyses have demonstrated the patriarchal nature of such a definition, based as it is on the notion of a public–private division within society. In this dichotomy 'the public' includes employment, the state and politics and is essentially a male sphere whilst the family and the house-hold (female spheres) are 'private' and as such, non-political. The location of 'politics' within this view of the world doubly excludes women. First, they have been largely excluded from entering those state institutions which wield public political power. Secondly, large areas of women's lives have been kept off the formal political agenda. More generally, gender relations and women's economic and social subordination have been relegated to the supposedly natural private sphere. This definition of 'politics' also excludes many other areas of activity from the political sphere – for instance, housing struggles or self-help organisations.

Recent years, however, have seen a shift in the popular definition of what is political. It is now widely, if not universally, accepted that many diverse forms of activity can be considered as political, not merely those activities which relate to electoral and state politics. What is considered 'political' has been extended, largely, it should be emphasised, as a result of deliberate struggles by feminists and others to redefine their experiences as political and not private, natural or personal. But this extension has not meant the unequivocal acceptance of the whole range of political activity into the legitimate mainstream political arena. A second distinction has emerged between 'formal' and 'informal' types of politics.

Broadly the term 'formal politics' is used to refer to political activity which is allied to the hierarchies of administrative power in local and central government – primarily political parties but also trades unions. Informal politics represent the forms of organisation such as voluntary and women's and ethnic organisations which are seen as being largely outside state structures and less linked to the exercise of executive power. Along with this general distinction a host of sometimes oppositional characteristics are associated with the two categories. For example, forms of organisation and participation are expected to differ with the 'formal' being hierarchical and rigid, concerned with economics, the Welfare State, the workplace and wages, and a mode of operation which utilises formal negotiations and meetings. The informal is thought to be more ad hoc and unstructured, concerned with issues such as childcare or self help schemes and characterised by direct action. It would be foolish to deny that different types of politics exist or that the characteristics described above apply more to some types and less to others. But we will argue that such characteristics do not appear rigidly in one sector or the other according to any pre-determined combination.

The terms 'formal' and 'informal' have created a new dichotomy with which to describe political activity...[and this] newer dichotomy does politicise gender rather than relegate gender relations to 'natural' and 'non political' spheres. However, it is also overlain with gender considerations. The characteristics of formal organisations are often associated with male forms of political organising. Informal politics on the other hand are frequently associated with, for example, the women's movement. To push the varied politics associated with the women's movement into an 'informal' category is, in our view, to deny the struggle which took place to redefine the personal into the political. There is a serious risk that the new dichotomy will be used in such a way as to repeat many of the tensions inherent in the 'public–private' dichotomy. In this way the challenge of women's and other movements is sidestepped by the re-imposition of separate spheres...

READING C

Valentine, G. 1989. The geography of women's fear. *Area*, 21, 385–390.

...It is well established in the sociology and criminology literatures of western Europe that women are the gender more fearful of crime and that this is related to women's sense of physical vulnerability to men, particularly to rape and sexual murder, and an awareness of the seriousness and horror of such an experience... However, little has been written about the geography of this fear...

The association of male violence with certain environmental contexts has a profound effect on many women's use of space. Every day most women in western societies negotiate public space alone. Many of their apparently 'taken for granted' choices of routes and destinations are the product of 'coping strategies' women adopt to stay safe... The predominant strategy adopted...is the avoidance of perceived 'dangerous places' at 'dangerous times'. By adopting such defensive tactics, women are pressurised into a restricted use and occupation of public space... A woman's ability to choose a coping strategy and therefore her consequent use and experience of public space is largely determined by her age, income and lifestyle...

Women assume that the location of male violence is unevenly distributed through space and time. In particular, women learn to perceive danger from strange men in public space despite the fact that statistics on rape and attack emphasise clearly that they are more at risk at home and from men they know...

The type of places in which Reading women anticipate themselves to be most at risk are...those where they perceive the behaviour of others, specifically men...to be unregulated...large open spaces which are frequently deserted: parks, woodland, wasteground, canals, rivers and countryside... [and]...closed spaces with limited exits where men may be concealed and able to attack women out of the visual range of others: subways, alleyways, multistorey car parks and empty railway carriages. Such opportunities for concealed attack are often exacerbated by bad lighting and ill considered and thoughtless building design and landscaping...

This inability of women to enjoy independence and freedom to move safely in public space is ...one of the pressures which encourages them to seek from one man protection from all, initially through having a boyfriend and later through cohabitation...

Box 3.4   Dualisms

A dualism is a particular structure of meaning in which one element is defined only in relation to another or others. Dualisms thus usually involve pairs, binaries and dichotomies, but not all pairs, binaries and dichotomies are dualisms. What makes dualisms distinctive is that one of the terms provides a 'core', and it is in contrast to the core that the other term or terms are defined. Thus dualisms structure meaning as a relation between a core term A and subordinate term(s) not-A. Many geographers, for example, have remarked that efforts to define the postmodern city are often dualistic, since many are simply the opposite of whatever criteria are used to define the modern city (Pile and Rose, 1992). The reason some feminist geographers have paid attention to dualistic ways of constructing knowledge is that dualisms are very often gendered and hierarchised, so that the core term A is masculinised and prioritised, and the subordinate term(s) not-A are feminised (Rose, 1993; Massey, 1994). This particular construction is also sexualised as heterosexual since it constructs difference as existing only between masculine and feminine; and postcolonial accounts of the colonised as the colonisers' Other are also describing a dualistic relation.

References

MASSEY, D. 1994. *Space, Place and Gender.* Cambridge: Polity, pp. 255–270.

PILE, S. and ROSE, G. 1992. All or nothing: politics and critique in the modernism–postmodernism debate. *Environment and Planning D: Society and Space*, 10, 123–136.

ROSE, G. 1993. *Feminism and Geography: The Limits of Geographical Knowledge.* Cambridge: Polity, pp. 73–77.

CHAPTER FOUR

Methods and methodologies in feminist geographies: politics, practice and power

CLARE MADGE, PARVATI RAGHURAM, TRACEY SKELTON,
KATIE WILLIS AND JENNY WILLIAMS

4.1 Introduction

The previous chapters have demonstrated that there is a vibrant debate surrounding theoretical issues in feminist geography. In particular, the themes of diversity and difference between feminist geographers, and approaches to their study of geography, have been highlighted. Here we continue the exploration of the contested nature of feminist knowledge through an examination of the methods and methodologies that feminist geographers use to inform their work.

Feminist geography, like feminist theory, is involved in challenging academic orthodoxies about how research is undertaken. This challenge is three-fold. First, it involves rethinking the categories, definitions and concepts used to formulate theories within the geographical discipline. Second, it involves examining the methods (and the theories underlying them) used for exploring defined problems. Third, it involves considering the process of selecting problems deemed to be significant for geographical enquiry, in particular through using gender (with all its multiple meanings and intersections with other aspects of social identity) as a key analytical category in studies of space/place, environment and landscape. This chapter explores the second part of the challenge, examining feminist interventions into methods and methodologies, highlighting the political motivation, the practical outcomes and the power relations involved in feminist research.

In Boxes 4.1, 4.2 and 4.3 you will find idealised general descriptions of the important terms related to research. In reality the meanings and uses of these terms overlap, are shifting in nature and are continually contested.

Box 4.1   Epistemology

Epistemology: The study of what constitutes knowledge. It is concerned with providing a philosophical grounding for deciding what kinds of knowledge are possible and how we can ensure that they are both adequate and legitimate (Maynard, 1994: 10). It is a theory of knowledge that informs our research strategies, e.g. positivism, behaviouralism.

Feminist epistemology: Feminism challenges traditional epistemologies of what are considered valid forms of knowledge. Feminist epistemology has redefined the knower, knowing and the known (Harding, 1987; Moss, 1993: 49). It questions notions of 'truth' and validates 'alternative' sources of knowledge, such as subjective experience. Feminist epistemology stresses the non-neutrality of the researcher and the power relations involved in the research process (D. Rose, 1993: 58). It also contests boundaries between 'fieldwork' and everyday life, arguing that we are always in the 'field' (Katz, 1994: 67).

Box 4.2   Methodology

Methodology: The theory and analysis of research procedures (Eyles, 1993: 50). e.g. quantitative, qualitative.

Feminist methodology: A methodology where links are forged between knowing and doing. Feminist methodology is committed to challenging oppressive aspects of socially constructed gender relations (whether these act alone or in conjunction with other oppressive relations based on race, class, sexuality, etc.). It recognises the social relations of research and has emancipatory goals for all those involved in the research process, leading to social change. Feminist methodology aims for mutual understanding and learning about the meaningful differences between the researcher and the people who are the subject of research with respect to structures of domination (Katz, 1994: 70).

Box 4.3   Method and practice

Method: A technique for gathering evidence (Eyles, 1993: 50), e.g. questionnaire survey, participant observation, in-depth interviews.

Feminist method: Research methods that are consistent with feminist goals. These may include quantitative (e.g. questionnaires and statistical analysis) and qualitative (e.g. in-depth interviews and textual analysis) methods. It is the theoretical orientation that guides the conceptual framing of the research, the questions asked, the application of methods and the interpretation of results that create feminist enquiry (Dyck, 1993: 53).

Feminist research practice: Encompasses all aspects of the feminist research process, i.e. the questioning of what is knowledge, the methodology involved and the actual techniques used to create that knowledge (Kelly *et al.*, 1994). While a differentiation is made between method, methodology and epistemology, the way in which a method is used will be affected by the epistemological and methodological perspective. Thus, in reality, these three terms are interconnected. The term 'research practice' may therefore be used to encompass all three terms and to highlight their interrelationships. When the explicit intention behind the research is feminist, then the term 'feminist research practice' may be used.

One of the great contributions feminism has made to the discourse on methodology has been its critique of positivist empiricism and non-feminist research methodologies. Such approaches were secure in their belief that they were the 'correct' way to conduct research. Positivist methods divide object and subject to ensure 'all important' objectivity. The researcher is created as omnipotent, expert, detached, neutral, free from bias and personality – 'qualities usually only attributed to angels' (McDowell, 1992: 405, quoting Grosz, 1986: 199). There is no opportunity for the 'personal' to intervene and complicate the research process, for positivist methods do not allow the 'personal' any room (England, 1994: 81). However, feminism disrupted such notions as the following quote illustrates: 'the western industrial scientific approach [positivism] values the orderly, rational, quantifiable, predictable, abstract, and theoretical: feminism spat in its eye' (Stanley and Wise, 1993: 66, quoted by England, 1994: 81). What feminism showed was that the world was not orderly and rational but full of contradictions and complexities. Increasingly geographers have attempted to investigate these contradictions through qualitative methods, realising that the 'personal' affects the way in which we do research: it influences the questions we ask, the ways in which we interpret answers to those questions, and what we do with our research results. It must, however, be stressed that many feminists would argue that the 'personal' also influences the quantitative research process, although this is often hidden behind a veneer of scientific neutrality. The shift to more qualitative research methods was therefore partly promoted by the feminist critique of positivism.

Despite this critique of positivism it is important to remember that feminist geographers have always used a range of different methods and methodologies in order to build feminist knowledges. Here are some examples: they have worked with huge data sets, such as census data, subjecting them to detailed analyses; large questionnaire surveys have been carried out which have allowed patterns and trends to be statistically identified; material/physical spaces have been investigated to examine the processes which produce gendered patterns of use; structured interviews have been used to find out what women (and men) think about particular issues; long and in-depth interviews have been conducted to gain senses of women's experiences of place, space, landscape and environment; cultural texts such as films, television, music, paintings have been interpreted for their gendered and gendering qualities; and feminist geographers have researched through 'thinking', developing theoretical and conceptual ideas through reading and dialogue. Feminist geographers have advocated using already existing research methods and methodologies, and have also insisted that the production of feminist knowledges requires new methods and methodologies.

A common thread binding much of this methodological diversity together, however, is an effort to be critically aware of the social contexts and consequences in which research takes place. The research process is always social: it involves academic debates and protocol; it recognises the complex relations between people carrying out research and the informants who are often called the researched; it considers the social contexts in which the research is made available for several audiences. Research methodologies are never considered by feminists as abstract processes of knowledge seeking. Rather, the social identities and power relations in which research is embedded, like any other social action, are given careful consideration in an effort to construct research which will contribute to the feminist project in all its diversity.

Chapter Three pointed out that there are different emphases in the ways in which feminist geographers understand the gendering of social difference. For some feminist geographers, gender is the most important category of feminist analysis, and gender for them refers to women as a group. Other feminist geographers argue that gender is inflected with other social relations such as class, race and sexuality. Yet other feminist geographers argue that gender is a category feminists should qualify and even critique, since social differences are so complex and we need to start thinking about difference in new kinds of ways. All these theoretical arguments intersect with discussion about which methods and methodological approaches to use when conducting research. The intersection is complex because it is not possible to list a method which is used for investigating a particular theoretical approach. There is much more diversity in the ways in which feminist methodologies are used. Feminists who have a similar theoretical approach may use very different methods; feminists with different theoretical approaches may find that they use the same methods for their research.

This chapter will provide you with an understanding of the diversity of methods and methodologies used by feminist geographers. The chapter is divided into four main sections. The first is the 'Introduction' which you have just read, and the second investigates conceptual themes in feminist research practice. In the third section we identify some key features which characterise feminist methodologies. The chapter closes with our conclusions. Throughout the text you will find activities and key readings to enhance your understandings of the issues raised.

4.2   Feminist research practice: conceptual themes

There have been ongoing debates regarding methodology in both non-feminist human geography (Keith, 1992; Pile, 1991) and feminist research in other social science disciplines (Reinharz, 1992). Human geographers' debates on methodology have been influenced by ideas from other disciplines such as psychology, sociology, anthropology and literary analysis. Thus in places the chapter explicitly considers feminist geography while in other places the discussion focuses on the interdisciplinary nature of feminist research. Additionally while we use the phrase 'feminist geography research practice', it must be realised that there is no *one* practice. Research methods employed by feminist researchers vary with the aims and location of the project, differ according to the intended audience or recipient of the finished research product, and alter according to the researcher's own interests, beliefs and positionality. Thus, feminist geographical research practice is both multi-stranded and complex.

This section investigates feminism's challenge to research orthodoxy by examining key themes that have been seen as important by feminist geographers. In some ways this is a difficult task because while there is undoubtedly a historical framework to feminist geography research, and the methods and methodologies used, it is important to point out that this does not imply a progressivist history. We cannot say that at certain points in time feminist geography used only one type of method, because different methods can be used to address the same questions and the same methods can be used to address different questions. There are not, therefore, stages of feminist geography research which progress through time; the processes relating to research, and the ways in which we conduct research, are much more complex than that. However, we do need to take you through some of the context relating to issues of politics, practice and power which informed, and continue to inform, contemporary feminist geography research. In many ways the best way to write such a context would be by using overlapping and interlocking circles but the printed page does not lend itself to such a representation. Consequently we take you through key conceptual themes which have been important in the ongoing debate relating to feminist geography research methods and methodologies.

Concepts in research: quantitative and qualitative techniques?

Feminist geographers have used both quantitative and qualitative methods since the inception of feminist geography, and they have tussled with the relative merits and demerits of each approach. Indeed, in the mid-1990s feminist geographers were again debating the use of both kinds of methods (see, for example, the journal *The Professional Geographer* suggested in the Further Reading at the end of this chapter). In this section we introduce you to what has become a highly complex debate which is not structured simply around quantitative versus qualitative methodological approaches but rather engages with criticisms of both types of method, the inadequacies of each approach and the political and ethical issues at play whichever approach is adopted.

Quantitative

'Quantitative', as its name implies, gathers data in order to quantify them in some way. Once the data have been turned into numbers or codes, they are then analysed mathematically, often statistically. They can be explored using relatively simple measures such as frequency counts, percentages or cross-tabulations, or by using more complex statistical techniques such as calculations of correlations, regressions or clusters. Quantitative methods have been commonly used in feminist geography research, and some examples are given below.

Feminist geographers have shown that geographical research has focused on white, bourgeois men's experiences and has ignored the experiences of women (Mackenzie *et al.*, 1980; Tivers, 1978). Questions about women have been added to the existing framework of geography as female geographers aim to make women visible through their research (e.g. Zelinsky *et al.*, 1982). A number of studies use census data and surveys to highlight labour force participation rates of women and so dismantle the image of the 'non-working' woman. For example, within agriculture women are found to be employed in a range of occupations from agricultural wage earners to independent small farmers (Momsen and Townsend, 1987). Other studies use time-use budgets and questionnaire surveys to complement statistical analysis, thus showing women's contribution to productive labour, family life and community work (Deere and León de Leal, 1982; Dixon, 1978). Archival material is also scrutinised to make women's multiple roles visible. Historical research, for example, shows the ways in which archival sources and archival indexing systems 'silence' women's histories. 'Official' sources of data are shown to contain a male bias. To overcome this bias, archival sources are contextualised and read in conjunction with other sources.

Although some feminist geographers continue to work with quantitative methods which are appropriate for the questions they ask, there are feminist geography criticisms of these methods. Initial criticism of quantitative methodologies revolved around three main concerns. First, the presence of pre-existing categories which were based on male experiences meant that the topics selected and the data collected for study were dependent on these pre-determined variables. For example, feminist geography research focusing on women and employment found that most censuses defined 'work' as wage work (Anker, 1983; Benería, 1981). Unpaid work by farm wives, work on subsistence farms and domestic work such as the processing of farm produce or child care, which take up much of women's time, were ignored. Thus feminists contested the accuracy of pre-existing categories used to describe and interpret women's experiences. Feminists argued for the need to ask questions relevant to women, to ask women these questions (rather than asking men questions about women), and to use women to ask these questions (as it was thought that women were more likely to reveal the true reality of their lives to other women rather than to men).

The second main criticism of quantitative methodologies revolved around the role of the researcher. The hierarchical and exploitative nature of the fieldwork experience where the researcher, in pursuit of an 'objective truth', was engaged in extracting information from informants was questioned. Feminists who had been instructed to maintain objectivity (i.e. to keep their distance from, and avoid revealing their own attitudes to, the 'researched') felt dissatisfied with the nature of their fieldwork experience. The third criticism centred on the need for contextualisation of data. Sensitivity to differences in language, behaviour and actions of different groups of women could only be obtained through in-depth qualitative research. Quantitative techniques amassed a number of atomistic facts, but the ways in which these interacted to determine a woman's lived experience were lost. Quantitative data thus appeared to be 'disembodied' as the methods used abstracted information, rather than reconstructing experience in a holistic way. For example, questionnaires on women's work missed out the way in which women often did several tasks at the same time. Women may be minding children, cooking and cleaning grain for the market simultaneously, but questionnaire categories often recorded only one task, thus artificially reducing the complexity of women's lives. It was also increasingly realised that women's lives were played out in conjunction with, and in opposition to, those of men, not separately from them.

Qualitative
Feminist criticisms (including those of some feminist geographers) of quantitative research have led to a refocus on other research methodologies, primarily qualitative, as well as modifications of existing methods to meet feminist objectives. 'Qualitative' techniques aim to explore the processes producing a particular event and to promote detailed understanding of socio-spatial experiences. They are intensive research methods, for example in-depth interviews and participant observation, often based on detailed case studies rather than large-scale data sets. They aim to understand the causes and the particular characteristics of the case study that is being researched. Qualitative methods offer interpretations of causal processes that have wide conceptual relevance. We present some examples of this below.

Some feminists feel that feminist research questions should generate from the personal experiences of women, from a female standpoint. Thus some research has been reoriented to explore and validate women's own views of their world. Small-scale, intensive studies using face-to-face research techniques conducted by women have been seen to be more appropriate to some feminist work, partly because they draw upon women's (purported) ability to listen, to empathise, and to validate personal experiences as part of the research process (McDowell, 1992; Oakley, 1990). Some traditional methods, such as the interview technique, have been transformed 'by advocating conscious partiality, a non-hierarchical relationship and an interactive research process' (Bergen, 1993: 201). The social relations of research are acknowledged which includes a recognition of different subject positions. The research process is thus a dialogic process jointly shaped by the researcher and the 'researched'. Feminists, therefore, have tried to develop a more humane, less exploitative relationship between researcher and 'researched' in which explicit consideration of the relevance of the research is made.

An example of such an approach where research is conducted as conversation rather than interrogation is consciousness-raising, where the informants discuss, understand and come to terms with their own personal experiences. Some of those involved in the research become empowered as the seeds for activism, for example self-help groups, have been sown. Maguire (1987), for example, based her doctoral dissertation on research which led to her setting up rape crisis centres, in what she has called 'action research'. For some feminists all research should be activist (Mies, 1991) but action research is not without problems. For example, Mies (1983) discovered that not all women who had been taken into a women's refuge/safe home wanted to be politicised through (the researcher's) consciousness-raising projects which tried to explain their position as a manifestation of patriarchal power. Women with whom one is working may not share the (same) feminist agenda and the question must be asked whether the researcher is justified in imposing their agenda on the 'researched'.

Overcoming the quantitative/qualitative binary
Recently there has been a recognition of the need to overcome the dichotomies between quantitative and qualitative research methodologies and methods as it is realised that the two are not entirely divorced from each other and are better seen as being on a continuum (see Hodge, 1995; Lawson, 1995; McLafferty, 1995; Mattingly and Falconer-Al-Hindi, 1995; Moss, 1995; Rocheleau, 1995). Indeed, it is quite possible for a single study to use both methods: quantitative methods to search for interesting patterns and trends, and qualitative methods to aid in the understanding of those trends. This is evident in the passage below in which Katie Willis describes the multiple methods she used in her research project which investigated the role of social networks and women's work in the household economy in urban Mexico:

> When deciding on a research method I considered a number of factors including: my limited time and money; my 'positionality'; the communities I would be working with; and what I was going to do with the data. This range of factors led me to adopt a range of techniques: a questionnaire survey to three hundred women; twenty semi-structured interviews; and participant observation. I went through the twenty-five page questionnaire with all three hundred sampled women and filled in their responses. This was partly because of illiteracy among some of the women, but was also because I wanted the data-gathering to be more than a one-way process. By spending time with every respondent I was able to explain my work and answer questions. It also meant that I received women's responses. With self-completion questionnaires there would be the possibility of men replying 'on behalf' of their partners, or of non-completion because of male disapproval. The questionnaire survey was used as a way of obtaining background information about three urban districts. I collected details about household composition, employment, housing quality and education. The lack of census tract data meant that this type of material could not be acquired through other channels. Following this initial survey I was able to provide the local residents' committees with some basic statistics about their communities. This reciprocal relationship was a very important part of my research as I wanted to do more than 'take my data and run'. (Willis, 1995)

Willis used a range of research methods. The questionnaire survey, which would be classified as quantitative, allowed her to develop a broad sense of what patterns might typify aspects of female life. The semi-structured interviews and the participant observation, usually classified as qualitative, allowed her to establish a more subtle and detailed understanding of women's roles and *their* interpretation of their gender roles and the questions of power at play for them within the household. She therefore grew to conceptualise these Mexican women's lives in all their diversity and complexity, and no doubt through these interactive methods the women came to understand a great deal about Katie Willis herself.

It has also become increasingly recognised that the sensitivity of qualitative research methods does not necessarily preclude exploitation, for those who are researched are still in danger of being manipulated and exposed (Stacey, 1988). Where the researcher engages in the life of the research community, the researcher's multiple identity as friend/researcher can become exploitative. For example, informants may disclose information to 'the friend' which they would rather not reveal to 'the researcher'. These issues are discussed in the extract below which was written by Clare Madge after returning from forestry-based fieldwork in The Gambia:

> My research in The Gambia employed a plurality of research methods, but participant observation played an important role. Particularly prominent to participatory research is the idea that a researcher can maintain a 'delicate balance of distance and closeness' (Shaffir and Stebbins 1991: 143), being at the same time an 'insider' and an 'outsider'. Participatory research suggests we should become involved in the 'lifeworld' of the research population, build up personal relationships, become friends. Yet the analysis, writing up and dissemination of information often forces us to detach ourselves, switch back to 'Western mode' to produce texts and develop 'distance' to use information gained through friendships. In other words, to become a stranger. In reality I found this delicate balance laden with complications. How can one be both a friend and a stranger? Indeed, as Ian Jarvie (1982: 68) suggests, to play the roles of friend and stranger with integrity, while trying to combine them is impossible. An example will best illustrate this point.
>
> During my year's stay in The Gambia I learnt much 'privileged information' through my personal relationships with individuals, informally chatting or through daily participation. However, after becoming a friend, I did not feel that I could suddenly become the detached stranger on my return to England and use such information for my academic advancement. For example, I learnt much privileged information about the use of herbal medicines for 'women's' complaints, but although one aspect of my study was the role of herbal medicines to rural Gambians, I did not use the information about women's herbal medicines in my thesis. To do so would have been to betray the trust of my friends, as in this context knowledge is linked to power; I may have disempowered them through the use of such information (I was sending the villagers a copy of my thesis so anyone could have gained access to that medicinal knowledge).
>
> This example renders suspect the validity of the dichotomous divisions of friend and stranger. Rather I suggest that participant observation involves playing out a multiplicity of changing roles during the course of the research. These roles, which are sometimes complementary, sometimes clashing, and which are contingent on our positionality, will influence the data given/gained and our subsequent interpretations. In other words, they will influence what we produce as knowledge. Personal relationships with people will influence the ethical decisions we

make regarding what we create as knowledge. Power, ethics and knowledge are interconnected. (Madge, 1994: 95–96)

Read the extract above and debate the following questions. Should the researcher have included the 'privileged information' in her thesis? Are qualitative techniques more or less likely to be exploitative than quantitative techniques? Can you think of any examples from your own research of how power, ethics and knowledge are interconnected?

While the use of qualitative methods by feminists is becoming increasingly problematised, there is a parallel appreciation of the advantages of employing quantitative methods in feminist research (Jayaratne, 1993; Pugh, 1990). One advantage of quantitative research is that large quantitative surveys can show the incidence/prevalence of particular phenomena and can force these issues onto the public agenda. Surveys have power to change public opinion in a way that a limited number of in-depth interviews may not; 'hard' data obtained using 'objective' quantitative techniques may be more acceptable to policy makers. One example is the way that the results of domestic violence surveys have forced the media to consider this widespread but little addressed problem (Kelly, 1988). A 'traditional' research method may therefore be used to further progressive change to promote the feminist project (Jayaratne and Stewart, 1991: 100). Qualitative methods have often been criticised as value-laden and subjective, and feminist research undervalued because of its association with such 'soft data'. However, many feminist researchers argue that quantitative research methods also come from subject positions. The design and content of questionnaire surveys, usually considered unbiased, are, in fact, a product of the subjectivity of the researcher.

A second advantage of quantitative research is that it can be used to test and invert old androgenic theories and practices. Just as some qualitative techniques have been modified, quantitative research techniques may also be modified to meet feminist objectives. Questionnaires may be sensitively designed so as to challenge and contest masculinist ways of thinking/seeing/knowing. Finally, the anonymity of individuals and the distance between the researcher and the researched which quantitative research affords may be crucial when exploring particularly sensitive topics (Kelly, 1990; however, see also Bergen, 1993 for the opposite viewpoint). In the extract below Jenny Williams discusses how some women involved in a research project on post-natal depression (PND) in the Wirral, northwest England, preferred the use of semi-structured questionnaires rather than informal interviews:

Keen to undertake qualitative research through informal interviews, thus affording sensitivity to women confronting a painful issue, I had rejected any notions of producing prescriptive questionnaires. However, during the project, several women telephoned me with similar requests. Explaining they wanted to participate, but their experiences were too distressing to verbalise, I was asked to compile a written questionnaire for them to complete. Consequently I revised my original plans and designed a semi-structured questionnaire based on open-ended questions. This provided a source of rich data in a way which was neither too disturbing nor embarrassing for the women. (J. Williams, 1994)

Feminist geography methods and working with difference

As feminist research focuses on differences *between* women, another facet of the debate around methodologies and methods has focused on the ways in which such differences can be researched (e.g. Gregson and Lowe, 1994; Johnson, 1989; Valentine, 1993a). As different women have distinct personal experiences, their research agenda (or what they consider important and in need of study) may be specific to these experiences, and research methods must reflect these varied agendas. Black women, for example, have criticised white women for making universalist assumptions about women and gender relations based on their own specific white experiences, therefore treating gender as the primary axis of social differentiation and grafting other axes such as ethnicity (or class, or sexuality) on to the category gender. Black feminists have also been at the forefront of the debate regarding the power relations involved in the research process, asking questions about the relationship between the researcher and the research community, such as whether white women can research black women (Maynard and Purvis, 1994) and whether it is ethical for women from the 'First World' to undertake research on women from the 'Third World' (Mohanty, 1991). These criticisms have also focused attention on the ways in which research is organised and funded, how research projects on certain issues may receive financial support while others do not, so undermining the concept of value-free research and stressing the need to locate each research question within its political context.

Recent feminist theory has therefore tried to destabilise the unitary category 'women' and in response feminist researchers have experimented with new research methods. Life histories/stories have been employed to allow voices often silenced by dominant discourses to be heard. By documenting the voices of 'silenced' people, history is rewritten in a more holistic way. For example, Janet Townsend and her collaborators used life histories to discover women's experiences of the colonisation of the Mexican rainforest (Townsend *et al.*, 1995). The chosen research methods allowed the researchers to alter the research agenda by listening to, analysing and reporting the women's most pressing concerns regarding their perceived lack of control over their own sexuality, child-bearing and the incidence of male violence. Longitudinal studies are therefore valuable because the research is not fixed in time but reflects the dynamism of people's experiences, so enabling temporal differences between women and men to be highlighted.

The use of life histories has been complemented by research which analyses dominant texts and reinterprets them with a feminist objective. Textual analysis interprets text as an image of reality. It has been used to uncover ideologies permeating a piece of text. For example, in Chapter Six the authors discuss various interpretations of Helen Allingham's paintings of English cottages and gardens in the nineteenth and twentieth centuries. The paintings (and their absences) are thus 'read' as texts which reveal information about gender roles and social relations of power based on gender, class and sexuality.

Speech and text are the most common means of communication, but the potential of other media to disrupt dominant discourses on gender and to destabilise the category 'woman' are increasingly being harnessed. We represent ourselves and the world around us in different ways, for example through artefacts, pictures, television, and video. Tracey Skelton (1995a) has analysed media coverage of a Jamaican ragga song in the gay and mainstream press to show the way in which these two media have emphasised sexual and racial stereotypes respectively. The study facilitates a more nuanced understanding of the intersection of race, sexuality and gender in the creation of social identities, as highlighted by the extract below:

> Trying to chart the resistance to the homophobia of ragga is, as with any process of resistance, complex... When such resistance appears at first sight to be white gay resistance against black Jamaican culture then the complexities of race and sexuality further complicate the issue. Such a dualism is inevitably going to be one which informs mainstream media representation... Their analysis will suggest that all the victims are white gays and all the antagonists are black; and that all black people are homophobic... Closer inspection of the resistance shows that it does not favour lines of race but rather of sexuality. In all presentations of the debate within the gay press the emphasis has been that this is a straight/gay problem not a white/black problem... It is the mainstream press... which entering the debate almost a year after it was begun in the gay press, attempt to establish a binary divide based on race. (Skelton, 1995a: 268)

ACTIVITY

Select a recent space-specific event which has received media coverage and consider the conflicting ways in which media represent the same event. Pay particular attention to issues of gender, race, sexuality, nationality, etc. For example, you could consider the UN Conference on Women, Beijing, China, 1995; the Bradford Riots, UK, 1995; the O.J. Simpson Trial, Los Angeles, USA, 1995; or the 'beef crisis' between Britain and the rest of the European Union, 1996.

Geographical research has, to date, privileged the visual, but geographers have been exploring other ways of knowing. Susan Smith (1994) has called for sound, especially music, to be analysed as an inherent part of the social landscape. Feminist geographical research has begun to explore these soundscapes. Skelton (1995b), for example, investigates the notion of femininity in the context of its television representation. She uses video material, secondary sources and her own experiences of fieldwork in the Caribbean to provide us with her reading of the way in which Jamaican women have used ragga to create the space to perform new femininities and sexual identities in the 1990s. These new methods can be used to explore women's identification of their own gender identities and to destabilise the idea of an essential gender category, thus giving a more realistic picture of the realities of women's lives, and challenging false representations based on gender, race, age and sexuality.

While recognition of multiple gender identities has disrupted the dualistic conceptualisation of gender, the politics of such feminism is less clear. J. K. Gibson-Graham (1994) offers some reflections on these issues:

> If we are to accept that there is no unity, centre or actuality to discover for women, what is feminist research about? How can we speak of our experiences as women? Can we still use women's experiences as resources for social analysis? Is it still possible to do research *for* women? How can we negotiate the multiple and decentred identities of women? (Gibson-Graham, 1994: 206, original emphasis)

In response to these questions Gibson-Graham tried to bridge the outsider/insider dichotomy between herself and the mining community in central Queensland, Australia, she was researching by employing members of the research community to assist in the project. Workshops for the research team set the scene for an exploration of the differences between women as well as initiating a process of 'partial identification' which is highlighted in the extract below:

> Interestingly, what emerged was not what I would see as identification around the shared experience of women (the recognition that *as women* we shared a common 'problem'). In fact personal differences in gender experience widened on many fronts. What took place was identification with respect to common problems of a very specific kind (ones that many women would not share) – living with a shift worker (or in the case of Joanne and myself, living with a self-employed partner who worked long and irregular hours, often including weekends); particular place-specific forms of male discipline; union, company and university reluctance to consider family life in industrial relations. (Gibson-Graham, 1994: 218, original emphasis)

This research opened up space for the emergence of new forms of alliances based on both alternative subject positions for gendered subjects, and of different forms of political resistance based on undermining the hegemony of binary gender discourse. This project illustrated ways out of the impasse which were initiated by the differentiation debate. Thus while acknowledging differences between women, certain research methods can allow for a 'space of identification' – or an area of common ground – between women which can be built upon as a source of resistance.

Conclusions

The discussion shows that feminist research methods and methodologies have both changed over time and attempted to use multiple modes: doing, hearing, seeing, writing and reading in different ways in order to uncover women's (multiple and varied) experiences. Feminism does not encourage methodological elitism, but instead promotes a plurality of methods where the choice of method depends on what is appropriate, comfortable or effective. Our own position is that feminist research can use any of the techniques mentioned above but the methods used must be appropriate to the research question being asked and should meet broad feminist goals. Hence no one particular type of research method is any more feminist than another (Staeheli and Lawson, 1994: 97).

We also believe that it is important that as researching gender becomes more 'fashionable', the political project, the feminist objectives of the research, must not be lost. Moreover, we would resist any new form of methodological elitism replacing the older forms that feminists have spent much time and effort in attempting to dismantle. We also challenge definitions of feminist research as 'research on, by and for women' as we believe that feminist research can explore both masculinities and femininities and the way in which they are produced simultaneously with other attributes of social identity, such as age, physical ability and location. Additionally, it has been argued by some that men may also conduct feminist research as long as they have feminist goals, but this is a contested arena as others believe that feminist research can only be conducted by those whose feminist consciousness comes from their personal experience, the experience of being a woman. The six authors of this chapter could not agree on this issue. Finally, it must be reiterated that women do not form one single constituency, so research that empowers one category of women may challenge the power base of other women. Women are affected by various forms of oppression, e.g. racism, homophobia, etc., but different women will challenge these issues differently under the broad political umbrella(s) of feminism(s).

ACTIVITY

This activity can form the basis of a tutorial discussion, a written assignment or a larger group debate. In the latter case ground rules should be established to ensure that the debate does not deteriorate into male/female, inferiority/superiority rhetoric. Preparation should involve careful re-reading of this chapter and wider reading of key articles/books on methodologies, feminism and gender. Below are several questions which participants need to think through and develop arguments around based on their wider reading:

1. What do you understand by the term feminist research? (Students will need a working definition of this before they can begin to discuss the role of the gender of the researcher. Try to be precise about the type of feminist research that is being conducted because for some types of feminist research gender may be important while for other types it may not.)
2. Do you have to share the identity of the researched community in order to share their political platform? For example, in a study on domestic violence, must you have experienced domestic violence as either a victim or a perpetrator in order to conduct research on this topic and to really understand the issues? In a study on racism in rural areas must you have experienced racism as a result of your social identity in order to study this topic?
3. We all have numerous identity positions, so is it possible to be an 'insider' in the research process? How can we decide which identity position is relevant to the research and which is not?

4. Is men conducting feminist research another way for men to speak for women and to retain power over them?

5. If knowledge is subjective and experiential (based on personal experience) can there ever be a basis for men conducting feminist research?

SUMMARY

- Feminist geography research practice has changed over time and continues to change as feminist epistemologies and geographical thinking change. Feminist geography methods and methodologies interact with feminist geography theoretical and conceptual thinking; each informs and influences the other in a dynamic relationship.
- Feminist geographers use a multiplicity of methods to inform their research, e.g. questionnaire surveys, interviews, textual analysis, media analysis.
- Different research projects will require the use of different methods, but the methods used must be appropriate to the research question being asked and must meet broad feminist goals.
- Feminist research practice is iterative and is committed to challenging oppressive aspects of socially constructed gender identities.

4.3   Characteristics of feminist geography methodologies

Feminist geography has been influenced by feminist theory. Feminist geography has also been in dialogue with other methodological approaches, such as humanistic, anthropological and interpretative approaches. Feminist geography is not therefore the only branch of geography to have critiqued positivist approaches, to have acknowledged the importance of positionality, to have adopted ethnographic methods and to make use of oral histories (Eyles, 1993). What remains distinctive about feminist research is that it places gender as a central (but contested) analytical category and that it insists that gender inevitably positions people in different ways.

In this section we identify four main characteristics which are important features of a feminist geographical methodology. These characteristics are not, as we note above, exclusive to feminist geographers and not all feminist geography research will exhibit all these features. The characteristics are not discrete but rather overlap and interlink. Nor is there constant agreement among feminist geographers about the relative importance or efficacy of these characteristics. Indeed, one of the key features of feminist geographers' research is its engagement with the methodological debate which is not closed, complete or 'sorted', but is rather like the very experience of research itself – it is an iterative process which is open to critique, reflection and debate. There are many more questions than answers regarding feminist geographers' interventions into methodologies.

Ways of knowing

Feminists have criticised the masculinist ethnocentricity of existing knowledge. They have redefined knowledge as both experiential (based on identities, experiences, social worlds and analysis of power relations) and interpretative (based on different interpretations of these experiences). As different women have different experiences, feminist geographical research should include the experiences of different categories of women. It has, however, largely focused on women, rather than men, and sexuality has only recently been given a central place in geographical studies (see, however, Bell and Valentine, 1995; Jackson, 1994; Valentine, 1993a, 1993b, 1993c). Feminist geography has also neglected, with some exceptions, the 'ways of knowing' among black women, women of colour, and disabled women (Chouniard and Grant, 1995).

As geographers, the 'place' we go to investigate this unstable, unpredictable, and differentiated knowledge is often referred to as the 'field'. The 'field' is an innovative arena of methodological debate which has been studied by feminist geographers. For feminist geographers, the 'field' is not a neutral place into which we step; it has its own political, social, economic and cultural context which will change over time. The 'field' will mean different things to different geographers and will be interpreted in as many ways as there are researchers and research projects. What we find out, what we perceive as important, where we choose to research, what 'knowledge' we decide is valid, is determined by our identity positions – our age, religion, gender, ethnicity, cultural background, sexual orientation, spatio-temporal location and so forth. Our identity positions are partial and situated, thus the 'field' we create from these positions is also partial and situated. Each field/place/environment we enter for research has commonality and difference from any other 'field' we experience.

ACTIVITY

Read the chapter by Ifi Amadiume, 'The mouth that spoke a falsehood will later speak the truth: Going home to the field in Eastern Nigeria', in Bell, D., Caplan, P. and Karim, W.J. (eds), 1993, *Gendered Fields: Women, Men and Ethnography*, London: Routledge, pp. 182–198.

The chapter highlights the complexities of the concept of the 'field' through a discussion by a Nigerian woman who is studying abroad (in the UK), going 'home' to conduct fieldwork in Nigeria. Referring to the article, discuss the following questions in a tutorial group:

1. If you were to embark on a study, which 'field' would constitute 'home' and which would constitute 'elsewhere'? Why?
2. What might be the advantages and disadvantages of researching at 'home' or 'elsewhere'?
3. What political and ethical issues come into play if you choose to work 'at home'? Are these different from the issues associated with working 'elsewhere'? If they are different, why are they different?
4. Why do so many 'First World' geographers conduct research in the 'Third World'?

In the following abstract, Jenny Williams describes how she constructed the 'field' for her research on women's health:

> The research I undertook for my MA thesis in Human Rights and Equal Opportunities was driven by a personal–political motivation – investigating a topic that had a personal significance: post-natal depression (PND). Years after my own recovery, and in the privileged position of having the benefit of a higher education, I wanted not only to examine current theories explaining PND, but also to contextualise the phenomenon within patriarchal discourses about women's (mental) health. Unlike previous research which attempted to quantify PND by measuring hormonal change or detecting psychiatric imbalances, I wanted to explore women's own experiences both of their depressions and the 'treatments' offered. Rather than accepting typical masculinist scientific 'knowledge' I sought to present a 'view from the inside'. Deciding to utilise qualitative research methods to contemplate this emotive topic, it seemed appropriate to work with a small group of women over an extended period (one year) collecting detailed information, then presenting the data by analysing women's own words. (J. Williams, 1994)

The centrality of Williams' personal experiences in the choice of this research topic is clear, exemplifying the way the 'personal' is crucial to the construction of the 'field'. Working with a small group of women sharing a common experience of PND and talking about this shared experience to develop an understanding constitutes a very different field from that which would be constructed by a male researcher approaching PND from a traditional 'scientific' angle. Through critically examining the way her 'field' was constructed, Williams' approach thus challenged masculinist approaches to PND. This example shows how feminist methodology considers the fundamental concepts of knowledge, how it profoundly questions dominant ways of knowing; it does not presume the accuracy of existing masculinist or ethnocentric constructions of the lives of women and men, for unless these underlying concepts are questioned, then the 'wrong' questions will continue to be asked.

Ways of asking

In order to construct these new, differentiated, inclusive knowledges through the use of deconstructed categories, we have to 'ask'. However, just as the way in which we construct our 'fields' is not free from politics, values and choices, so too the way we 'ask' and what we 'ask' is bound up within relations of power. Also what we ask, who we ask, where we ask, and when we ask, inevitably depends upon certain ethical decisions which present multiple dilemmas during the research process. Once we acknowledge that we, the world and the people in it are varied, different and positioned, then questions of power and politics become an important feature of our 'ways of asking'.

In thinking through our positions as researchers we have to think about not only power and politics but also the ideas of 'insider/outsider' in our research. As you will see below, this latter concept is one on which feminist geographers have not reached full agreement. Feminist geographic research has, under the influence of feminism, come to recognise explicitly the power relations involved in the research process and to acknowledge that such power relations are complex and multifaceted. Power relations will take different forms in different places, at different times and between different people. The relations may be founded upon material advantage, colonial history, the stereotype of the 'other', and the positionality of the researcher: power relations thus vary both spatially and temporally and are constructed, and played out, at a variety of scales (from the bodily to the geopolitical). Such power relations come into play not only for feminist geographers considering research in the 'Third World', but also for those doing research 'at home'. England (1994), for example, talks of abandoning her research on lesbian communities in Toronto. As a straight, white, middle-class academic woman she began to ask herself some difficult and yet crucial questions before embarking on 'asking' her research subjects:

> Could I be accused of academic voyeurism? Am I trying to get on some cheap package tour of lesbianism in the hopes of gaining some fleeting understanding of, perhaps, the ultimate 'other' given that lesbians are not male, heterosexual, not always middle-class, often not white? In the midst of academic discourse on the problems of appropriating the voices of marginalised people and the perils of postcolonialism, I worried that I might be, albeit unintentionally, colonising lesbians in some kind of academic neoimperialism. (England, 1994: 84)

These are difficult questions but as feminist geographers it is essential that we ask them of ourselves, for increasingly they will be asked of us. What is probably harder to accept is that the issue of conducting ethical research does not involve a simple set of formulae; rather it entails thinking through the process of working and living with people and attempting to challenge unequal power relations upon which differences rest (Madge, 1994: 98). We may then decide, as England did, to abandon the project; we may have the option of changing the focus of the research; we may decide to acknowledge the difficulties and continue with the research but include the dilemmas as part and parcel of the research findings.

Ultimately the decision about the power relations we engage in and the questions we ask during our research have to be based upon our identity positions, our politics and our purpose in carrying out the research. What is essential for us as feminist geographers is that we are honest about our research, the ethical dilemmas it raises and the outcomes of the research process.

ACTIVITY

Read the chapter by Daphne Patai, 'US academics and Third World women: is ethical research possible?', in Gluck, S.B. and Patai, D. (eds), 1991, *Women's Words: The Feminist Practice of Oral History*, London: Routledge, pp. 137–154. In a tutorial group, debate whether it is possible for 'First World' academics to undertake ethical research in the 'Third World'.

Once we have recognised the imbalance of power between researcher and research community, we then have to work towards a research methodology that aims either to counter such imbalances, or at least to lessen them. Kobayashi makes the relationship between the recognition of the power relations and the methods we use clear:

> The politics of involvement, for feminist researchers, require research methods that recognise the relationship with others as one of (ideally) mutual concern and trust... methods that stress mutual respect and involvement, shared responsibility, valuing difference, and nonhierarchical ways of achieving ends are not simple or shallow gestures of accommodation, nor are they just an alternative methodology. Such methods [and the way they are used] define an approach to political change. They raise the question to that of 'Who speaks with whom?' This question occurs before we enter the field and remains with us as we engage our subjects across the space of social engagement. (Kobayashi, 1994: 76)

One of the concepts related to 'ways of asking', which is part of the consideration of power relations, is that of the position of the researcher as an insider and/or outsider, and whether either position can alleviate the relations of power operating during the research process. Feminists cannot agree on whether black women should be the ones to research black women, lesbians to research lesbians, disabled women to carry out research on disabled women, etc. For example, although black women are not a unified group, within certain societies they all have the common experience of racism and so that particular aspect of the power relation is lessened. However, it could be argued that because there are so few women from marginalised groups within the academy then marginalised women's knowledges will never be sought and recorded if only the marginalised can 'ask' the marginalised. There is also the consideration that doing work as an 'insider' leads everyone to be 'pigeonholed' into conducting research on the group(s) with which they identify. This of course returns us to the debates about essentialist understandings of difference (see Box 2.2 in Chapter 2).

Dyck (1993: 54–56) discusses the role of the researcher through two scenarios, insider and outsider research. She investigated two research projects based in the suburbs of Vancouver, one with Caucasian mothers and the other with Indo-Canadian mothers. In the first study Dyck claims that her insider status, a mother, white and a resident of the same suburb, gave her the basis on which to conduct the type of research she felt most appropriate – interactive interviews conducted in an intimate manner. In her second study Dyck recognised that she was an outsider; being a woman was not enough to endow insider status. Faced with problems of conducting the interviews, Dyck found Indo-Canadian research assistants from among her subject group to work with her, as they were 'insiders'. They carried out the interviews with the Indo-Canadian women, using debriefing sessions to reflect on how the interviews were progressing. Dyck then reflected on the way these two different positions necessarily affected the way the 'asking' was done and the ultimate findings of the research, concluding that in both studies the responses of the interviewees were understood because the researchers could contextualise the cultural meanings and social lives of the respondents.

The issue of insider/outsider is complex and has been examined more critically by Melissa Gilbert (1994: 92) who concluded that she was just as much
an outsider among low-waged women workers in Massachusetts as she would
have been had she carried out her research in the 'Third World'. Her experience and subsequent reflection on the research meant that she no longer felt
that the feminist methodological position of 'insider' was very much help;
rather she concluded that we are, by the very fact of being researchers, 'out-
siders'. As Staeheli and Lawson argue (1994: 98) we must recognise the
'pitfalls of essentialising women as feminists and "sisters" [and rather understand that] women are differently positioned in the webs of power relations
that structure our identity'. Gilbert does not call for an end to research but
rather for a more critical approach to feminist methodologies and a consideration of the power relations that remain.

So far the discussion assumes that the researcher has power over the
research participants. However, it is increasingly recognised that the
researcher is not always in a position of power and this power is also not
absolute. For example, the researcher's agenda may be set by the 'gatekeepers'
of knowledge who decide what will be funded by research committees
(Schoenberger, 1991, 1992). Additionally, although the researcher may be in a
position of material power with respect to the 'researched', she or he may also
be manipulated and controlled by the research population. Townsend *et al.'s*
(1995) research in the rainforests of Mexico reveals that the Mexican women
who were the subjects of their research had clear agendas which they were
determined to present in the final documents for their own personal and/or
political ends. Additionally, the research product is also determined by what
will be considered acceptable by journal editorial committees or publishers.
Finally, the author has no control over the ways in which the findings will be
read or interpreted by the audience. We close this section on 'ways of asking'
with a quotation from Kim England on just such relations of power:

> In general, relationships with the researched may be reciprocal, asymmetrical, or
> potentially exploitative; and the researcher can adopt a stance of intimidation,
> ingratiation, self-promotion, or supplication. Most feminists usually favour the
> role of supplicant, seeking reciprocal relationships based on empathy and
> mutual respect, often sharing their knowledge with those they research... [Sup-
> plication means that] the researcher explicitly acknowledges her/his reliance on
> the research subject to provide insight into the subtle nuances of meaning that
> structure and shape everyday lives. Fieldwork for the research-as-supplicant is
> predicated upon an unequivocal acceptance that the knowledge of the person
> being researched (at least regarding the particular questions being asked) is
> greater than that of the researcher. Essentially, the appeal of supplication lies in
> its potential for dealing with asymmetrical and potentially exploitative power
> relations by shifting a lot of power over to the researched. (England, 1994: 82)

Ways of interpreting

We have established that feminism and feminist geography are determined to
disrupt masculinist, ethnocentric and homophobic notions of knowledge and

to destabilise many of the categories that appeared fixed in the establishment of that knowledge, most notably the notion of 'woman' and 'gender'. We have shown that feminist methodology understands that there are power relations, ethical questions and dilemmas to face in the ways we choose to 'ask' in order to construct our 'knowledge'. A third key characteristic of feminist geography methodology concerns the ways in which we interpret the 'knowledge' we have acquired through our asking (and in many ways it is closely associated with our fourth characteristic, 'ways of writing').

A key factor in feminist methodological ways of interpreting is the legitimation which is given to subjective knowledge. This means feminists argue that people know as much (probably more) about their lives, and the meanings they live with, as do those who attempt to study them. Feminist methodology also insists that research findings are interpreted within the context of the cultural framework of the research community, within its own autonomous systems of values, behaviour, attitudes, sentiments and beliefs, in order to contextualise the data and overcome white ethnocentrism (Ladner, 1987: 80; D. Rose, 1993: 58). The veil, for example, has been interpreted as a symbol of the subordination of Muslim women by Fatima Mernissi (1987), while Homa Hoodfar (1991) considers it as a way of subverting restrictions on women's use of space after the resurgence of religious fundamentalism in Egypt. Similarly, courtesans in the 'kothas' (brothels) of Lucknow, India, use their veil as a way of denying men any 'free' pleasure in viewing them (Oldenburg, 1990). Hence, there are multiple meanings associated with the use of the veil and, as part of the recognition of the differences among women, there has to be a determination to understand the different interpretations offered by these women in relation to their own particular cultural context.

A central part of the process of interpretation is reflexivity. We need to think about our assumptions, our part in the research process and the ethical considerations we make during the research process (Dyck, 1993: 53). Dyck argues that reflexivity involves a particular consideration of gender in order to be aware of the gendered nature of social and geographical phenomena, but also to reflect upon the effect that the gender of the researcher may have had on the research. England (1994: 82) reminds us that research is a process, not just a product, and that part of this process is reflecting on, and learning from, past research experiences. We should evaluate our research critically and attempt to interpret it in its full cultural, social, political and economic context, otherwise we will be guilty of false representations. We have to allow the multiple interpretations that come through the research to be acknowledged and presented. The interpretations we have of our research have to be written in such a way that the diversity, complexity and contradictions are given a space.

Ways of writing

Feminist geography methodology expects that the positionality of the author(s) will be acknowledged in the writing of the research; not to do so would be to revert to positivist or masculinist assumptions about the distance and objectivity of the researcher. However, this needs to be treated with sensitivity in a political world. For example, for some feminist geographers establishing their identity position as a lesbian could be very difficult for them within their workplace, with their students and within the academy as a whole.

In order to disrupt the balance of power in the research process, maintain the importance of subjectivity established through interpretation, and break down the hierarchies between researcher, research community and audience, the writing up of research should create a space for the voices of the researched to be heard. There are several techniques for doing this. One is to use personal pronouns in the text to highlight the partial nature of the research, rather than hiding behind a distant third person which depersonalises the research process. Thus the researcher does not appear as an invisible, anonymous voice of authority, but as a real, historical individual with concrete, specific desires and interests (Harding, 1987: 9). Through these techniques the researcher acts as a pivot for involving the audience in the lives of the researched. A further technique is to use the respondents' own words and to integrate transcript material throughout the final written version.

In her presentations and her research on Jamaican women and ragga music in conferences and in written form, Tracey Skelton makes determined efforts to use the words of a range of Jamaican women as they discuss what ragga means to them in Isaac Julien's film *The Dark Side of Black* (1994). Skelton shows clips from the film in presentations and quotes the women directly in her article. The written article also contains drawings of some of the dance hall movements used by the women. Skelton justifies the use of own words in this way:

> I draw on this part of the film for description and direct quotation because it is relatively rare that we see and hear Jamaican women talking about Jamaican ragga and their experience of it. What Julien's film shows is that the women do not all agree and I think that this is important as it is clear that ragga means different things to different women but that there is an engagement between them and also with men in ragga... Through using comments from Jamaican women themselves this essay has tried to show the sexual power of women when they dance to ragga and the potential ragga has for the empowerment of women and for the creation of new assertive and autonomous femininities. (Skelton, 1995b: 101–102)

Alternatively the taped voices of the researched may be used in an oral presentation (unless anonymity is required). Anne Opie (1992) states that the intensity of the spoken word can convey meaning which mere words on the page cannot, and that this should be indicated in some way. She also points out that the emotional content of tone of the response can tell us a considerable amount about the feelings and interpretations of any particular informant. Opie argues that multiple voices should be incorporated into the research because this is invaluable both to theory and also to the empowerment of participants – all of them have something important to contribute.

In writing our research as feminist geographers we have to think about who our audience is and whether any particular form of 'writing' is better suited to the research project – i.e. the use of visual displays, the use of copies of the taped voices of the participants, formal reports, radio interviews, television programmes. We also have to consider the cultural, political, social and economic context in which the 'written' form of the research process is received. We have to consider whether our research will perpetuate damaging stereotypes (Gibson-Graham, 1994: 208), place the participants in some sort of danger (England, 1994: 84) or expose practices of politically powerless people to those who may use this knowledge to oppress them (Katz, 1994: 71). Consequently, at each stage of the 'writing' of our research we once again have to make ethical decisions and so maintain a firm editorial control. However, that can often place us in a contradictory position vis-à-vis another debate within feminist methodology about the involvement of participants in the writing up of the research.

Gilbert (1994) discusses some of the techniques that have been used to equalise the power relations between the researcher and the researched with respect to the final written product. One is to present any conflicts between the researcher and the research community in the final version. Another is to give co-authorship to the women with whom they worked. Gilbert herself has presented some of her analyses to the women who were involved in the project to gather their opinions on the research. This process of negotiation has ensured participants' involvement through every stage of the research and provided them with the opportunity to contribute to the final written report. Such collaborative ways of writing are not, however, without problems:

> Recently, the practice has developed, particularly among anthropologists [and feminist geographers], of giving a draft of a report to research participants and asking them to comment on its validity. The point of this practice is to realign the balance of power in the research relationship by minimising appropriation through a deliberate attempt to avoid misrepresentation and stereotype and by the expansion of the researcher's appreciation of the situation as a result of discussing and reworking the text with the participants. However, I am unclear how agreement over the final version is reached when there is more than one participant. If agreement cannot be reached, one course of action is that the contentious material be removed; another, that the interpretation of some of the participants is privileged. However, I believe the situation to be more complicated. Removal of the material does not permit it to be discussed; while the subordination of one reading raises several problems. (Opie, 1992: 62)

While on the surface this engagement between the researcher and the 'researched' at the final stage of the project seems to be a good way to continue to share the research process and disrupt the usual power relations of the 'writing up', in fact it carries with it other complications. Ultimately we have to acknowledge that as researchers we have the final editorial control over the project and it is in this arena that we must ensure that our feminist politics, those of the research community (and any conflicts between those) are reflected in the final written version, whatever form that may take.

Use the following questions as the basis of a tutorial discussion. What are the advantages and disadvantages of presenting our research in ways other than the written form? Referring to Opie's article, discuss why the research participants should speak for themselves. How can you let the researched 'speak for themselves' within your research project (e.g. quotations, photographs) and what are the issues at stake if you do?

Conclusions: ways forward

As the above discussion has shown, feminist geographers, in common with other feminist researchers, frequently do not agree on debates over methods and methodologies, nor do they pretend to have all the answers. In many ways the formal, published debate on methodologies within the sub-discipline is relatively recent and there remain many questions. This is part of what makes the debate on feminist methodologies so interesting because as researchers continue their feminist research they can contribute to the methodology debate with new ideas, new techniques and new political insights. One thing should be clear now and that is that part of a feminist research project is an awareness of its limitations, of its contradictions, of its non-universality, of the role of subjectivity, and an acknowledgement of positionality and the issues of power. Other key aspects of feminist projects are reflexivity within the research process, consideration of the relationships between the researcher and the 'researched', as well as a commitment to challenging oppressive aspects of social identity.

As a consequence the following are some of the questions which a number of feminist geographers are currently addressing. What about our relationship with the 'field'? Can we ever really leave it? Can we, or should we go back? Why do people agree to be involved in a project? What happens to those involved in the project when we leave? How do the respondents feel after the project is complete? What are their reflections on their involvement? Can we establish alternative fora for the discussion of our research other than teaching, conferences, journal articles, books? Can we create a space for some of those involved in the project to participate in its presentation and what issues are at stake if we do? Is the academy ready to relinquish control of knowledge and share ownership equally with our participants? How can feminist geographers further their research methods and methodologies to promote positive social change and to challenge unequal asymmetries of power so that their research has a practical active outcome?

SUMMARY

- There are four main features which characterise feminist geographers' methodologies: ways of knowing, ways of asking, ways of interpreting and ways of writing.
- Not all feminist research projects will contain all features, but feminists will strive to achieve these characteristics through the research process.

- Conducting feminist research is not easy – it involves active reflection on the research process, careful consideration of the relations established on the research project, a clear understanding of the feminist aims of the research project and how these may be achieved, a commitment to challenging oppressive aspects of social identity based on gender, racial and sexual, etc., inequality, and an awareness of the limitations of the research and the important role of subjectivity.

4.4   Conclusions

In this chapter we have discussed some aspects of the wide debate within feminist geography surrounding methods and methodologies. Feminist geography is continually asking questions about the way research is conducted, presented and used, and about the ways in which we, as feminists and geographers, use our research to construct knowledge and understanding of the spatial and social world. The chapter has shown that feminist debates within the academy have influenced geographical methods and methodologies and points towards the ways in which feminist geographers, with their sensitivity to the spatial variation in methods and methodologies, may influence broader debates in the future. This fluidity and interchange is also highlighted through the circular connection between theory, methods and research. Feminist theory influences the research carried out and the findings of such research then influence the theory. In this way, feminist research is highly dynamic and reflexive. Changes in methodologies and methods are part and parcel of this circular, critical engagement between feminist theory and feminist research. As this chapter has shown, the methodologies used by feminist geographers have changed as the theoretical questions they asked, and the research findings they gathered, altered.

The chapter has also highlighted some of the characteristics which are important in the definition of a feminist geography research project. As we made clear at the outset, not all projects will contain all the characteristics we discuss, but to work towards attaining these features is an important goal of feminist research. The chapter highlights that feminist research in geography is not neat, clearly defined or neutral. It is, by its very definition, messy and problematic – and we can speak from experience. In asking different questions from the mainstream and in allowing respondents a more active role in the research process, you may not get clear answers, but you will probably get much closer to the 'reality' of women's and men's lives. Carrying out and writing up research in a non-traditional way (for example, by not setting up a hypothesis and then testing it through the use of 'scientifically' tested sampling methods, by not following an 'approved' formula of 'objective' writing) will require a lot of commitment and a lot of 'discussion' with your project supervisors. However, if you feel that feminist methods and methodologies will allow you to conduct your research in the most appropriate way, there is much to be gained from their use: it is important that feminists critique and challenge 'mainstream' methods and methodolo-

gies in order to dispute the hegemony of patriarchal institutions and masculinist constructions of knowledge.

Through their research feminist geographers have altered the structure of
the subject dramatically. Feminist research methods and methodologies are
definitely on the 'map' in geography. With the use of this chapter you too can
contribute to the 'new' body of geographical knowledge which has been
established and can carry out feminist geography research projects. We hope
that you will enjoy doing these projects as much as we have enjoyed doing
ours. Some examples of such feminist research projects are contained in the
remaining chapters of the book.

Further reading

Below we provide references from the journal for the professional body of geographers in the USA, the American Association of Geographers. Their work
has been very influential in the debate on methodologies in feminist geography
and also demonstrates that the discussion around methodologies is on-going
and dynamic. The 1995 references are interesting because they show that there
is a growing dialogue between qualitative and quantitative research methodologies. The papers are all found in *The Professional Geographer*, volume 46,
part 1 for the 1994 references, and volume 47, part 1 for the 1995 articles.

England, K. 1994. Getting personal: reflexivity, positionality, and feminist
    research, 80–89.
Gilbert, M. 1994. The politics of location: doing feminist research at 'home',
    90–96.
Hodge, D.C. 1995. Introduction to 'Focus: should women count? The role of
    quantitative methodology in feminist geographic research', 426–427.
Katz, C. 1994. Playing the field: questions of fieldwork in geography, 67–72.
Kobayashi, A. 1994. Colouring the field: gender, 'race', and the politics of
    fieldwork, 73–80.
Lawson, V. 1995. The politics of difference: examining the quantitative/quali-
    tative dualism in post-structuralist feminist research, 449–458.
Mattingly, D.J. and Falconer-Al-Hindi, K. 1995. Should women count? A
    context for the debate, 427–437.
McLafferty, S.L. 1995. Counting for women, 436–442.
Moss, P. 1995. Embeddedness in practice, numbers in context: the politics of
    knowing and doing, 442–449.
Rocheleau, D. 1995. Maps, numbers, text, and context: mixing methods in
    feminist political ecology, 458–466.
Staeheli, L. and Lawson, V. 1994. A discussion of 'women in the field': the
    politics of feminist fieldwork, 96–102.

In and out of bounds and resisting boundaries: feminist geographies of space and place

NINA LAURIE, FIONA SMITH, SOPHIE BOWLBY, JO FOORD,
SARAH MONK, SARAH RADCLIFFE, JO ROWLANDS,
JANET TOWNSEND, LIZ YOUNG AND NICKY GREGSON

5.1 Introduction

The use of binary categories in everyday language, and in turn in the discipline of Geography, is so widespread, and their common sense meaning so embedded in our lives, that we often take them for granted. This, perhaps, is not surprising. Binary categories such as public/private and global/local appear to offer the possibility of describing and analysing the world in a way which appeals to ideas of stability, completeness and authority. They seem neat and are easily understood, in some part because they often draw on elements of the dualistic thinking embedded in the structures of Western thought (see Box 3.4). For geographers, however, one of the most important things about binary categories is the ways in which these relate to the central geographical concepts of space and place.

Binary categories often suggest the existence of discrete spaces, that is spaces associated with different types of activities. In this respect, then, they could be said to appeal to ideas of absolute space (Section 1.2), and they often depend on the drawing of sharp lines between the two halves within the binary category. These lines are like boundaries, or fences; they are put up between the two sides which comprise the binary category. So, for example, the public/private distinction works by establishing a boundary between the public and private sphere, and by drawing a line between public and private space. Recognising the importance of boundaries to binary categories is critical. Indeed, what this suggests for many geographers is that binary categories are grounded in relational, rather than absolute, views of space.

In Section 1.2 of Chapter 1, we saw that relational understandings of space conceptualise space as structured by social, economic, political and cultural aspects of social relations. 'Social phenomena and space [are] constituted out of social relations, [such] that the spatial is social relations "stretched out"' (Massey, 1994: 2). Most feminist geographers have worked and, as we show in this chapter, continue to work, with this notion of relational space. However, the particular social relations which feminists have mapped have varied. For some, patriarchy's spaces have been the primary focus (Pain, 1991; Seager and Olson, 1986). For most, though, it is the intersection of patriarchy with capitalism that produces the spatialities of most concern (see Box 3.2). The spatiality of capitalist patriarchy has been explored by many feminist geographers, often influenced by the socialist–feminist position elaborated by Suzanne Mackenzie and Damaris Rose (1983). Mackenzie and Rose (1983) argued that the social relations of capitalist patriarchy consisted of the relations of production – waged work – and of reproduction – unwaged domestic work. These two forms of labour are profoundly connected, for reproductive labour reproduces the waged labour force in a number of ways: through biological reproduction, by servicing everyday needs, and by maintaining a haven free from the stresses of paid work. Patriarchy genders these social relations such that men are expected to undertake waged, productive labour and women are expected to undertake unwaged, reproductive labour. Mackenzie and Rose then argued that this gendered division of labour created, and was recreated by, a complex and historically dynamic spatiality. For example, within the home, certain spaces became gendered as feminine because they were the sites of much reproductive labour: the kitchen, for example (see Section 6.2). Moreover, the home itself became a space of reproduction in the nineteenth century, when industrial capitalism was reorganising into factory-based production (although, as we shall see in this chapter, this process was uneven and never complete). Given these economic and social changes, by the end of the nineteenth century in Britain, 'women's waged role came to assume an unnatural and then a dangerous appearance' (Mackenzie and Rose, 1983: 168). Women's proper place was seen to be the home. And as this process developed, city spaces also began to be gendered; the separation of work from home produced suburbia as a space separate from waged work, and this space was also gendered as feminine as 'the legitimate sphere of women's activities was confined to the home and neighbourhood' (Mackenzie and Rose, 1983: 170; see also McDowell, 1983; Miller, 1983, 1991). Thus Mackenzie and Rose argue that production and reproduction, the related spheres of capitalist patriarchy, produce a specific spatiality because of a particular division of labour: a binary division of spaces into workplace and home, the spaces of productive and reproductive labour respectively. Currently, many feminist geographers are further refining this argument by considering the spatialities made by other kinds of social relations: sexuality (Valentine, 1993a) and 'race' (Preston et al., 1993; Peake, 1993), for example.

An important point to note about this feminist analysis of the spatialities of social relations, and one we will return to in the conclusion of this chapter, is that it suggests that spatialities are both made materially and also maintained by any number of ideas about appropriate (gendered) behaviour and values. Implicitly, then, these arguments about boundaries and space also refer to the notion of place as constructed by 'maps of meaning' (see Section 1.2).

Boundaries, then, are central to Western conceptual frameworks of looking at the world, and in turn are expressed materially in the way the world is organised. Perhaps one of the most graphic examples of such material boundaries can be seen in the continued use today of the terms 'First World' and

'Third World', even though these terms have outlived their original meanings, located as they were in the capitalist–communist opposition of the Cold War period. The continued use of these terms, both academically and in everyday language, testifies to the time, effort and resources which are deployed in maintaining and reinforcing the boundaries drawn between both sides of a dichotomy, and the implicit hierarchies which may exist within it. Such activities can be thought of as a kind of policing, and entail relations of power, resistance and struggle between different groups, and their contrasting (and conflicting) perspectives. In the contemporary period, however, the boundaries between public/private, global/local, First/Third World, East/West, and so on and so forth, appear to be shifting radically. Boundaries between dichotomous categories no longer appear stable, and questions are being asked about both the content of various binary categories and the fences drawn within them. Indeed, questions are even being asked as to whether things were ever as stable as they once seemed.

Such questioning is both intellectual and has political implications. Politically, different groups are attempting to realign binary categories in diverse directions: the New Right in Britain and the United States, for example, want a very different organisation of public and private spaces than do diverse feminist groups. Donna Haraway has called these contemporary debates around society and culture 'border wars' (Haraway, 1991), emphasising the contested and undecided nature of boundaries, spheres, spaces and subjects. Feminist geographers, however, have always problematised the dichotomous spatialities constructed by ideas about what people and places are appropriate for what kinds of labour. Mackenzie and Rose (1983), for example, noted that the home is, for many women, also a place of hard work; and that many women have historically worked, and continue to work, for wages, both inside and outside the home. Feminist geographers continue to critique the gendering of dichotomous categories used to create space and places, the boundaries which these set up, and the ways in which boundaries work to define some people (and practices) as being 'in bounds' while others are located 'out of bounds'.

Feminist explorations of the construction of dichotomous spatialities is a project directed at Geography also. Geography as an academic discipline tends to focus on the spheres of life which have historically been masculinised – the political and the economic – and also to take for granted some of the distinctions which follow from the spatiality analysed by Mackenzie and Rose, among others – the social is often kept distinct from the political and economic, for example. Feminist geographers thus have two tasks: both to problematise the taken-for-granted nature of such distinctions within Geography, and to explore the way gender is spatialised. Much of the latter work has been concerned to show that dichotomous ways of thinking about spatiality are misplaced, and that many women (and men) are themselves struggling to overcome such dichotomies. These tasks have been undertaken by feminist geographers using different approaches to gender (see Section 3.3). Sometimes, challenging the invisibility of feminised work and spaces means that feminist geographers are concerned to produce a 'geography of women'; but feminist geographers have also been concerned to consider the ways in which gender roles and relations can account for the complex dynamics of gendered spatiality.

Boundaries, then, often have gendered implications. In the remainder of Chapter Five we focus on two of the ways in which feminist geographers have explored these issues. These are:

- by *reclaiming* one side of a binary category, often the side which has been less valued by non-feminist geographical analyses, and by making previously invisible processes, patterns and experiences more visible; and
- by illustrating that the boundaries drawn between binary categories are frequently more *blurred* than the dichotomy constructs them as, very often because women and men struggle to resist them.

We do this in the context of debates over two of the binary pairs which have most exercised feminist geographers: home and work, and informal and formal. The chapter tackles the issue of home and work in Section 5.2. Firstly we look at the ways in which feminist geographers have reclaimed the home/work dichotomy, in order to show the gendered relations producing both. We show how feminist geographers have reclaimed women's contribution to both paid work in the so-called 'First', 'Second' and 'Third World' contexts, and the significance of frequently invisible home spaces. We then examine how feminist geographers have shown the boundaries drawn between home and work to be more blurred than they are conventionally assumed and shown to be. In Section 5.3 we turn to considering the formal/informal political dichotomy, focusing particularly on some of the ways in which work in this field reclaims the undervalued and invisible within the arena of the political. We show how women are often invisible in formal politics, discussing here both the gendered nature of the state and a case study of the female refugee. The section then moves on to consider the other side of the formal/informal dichotomy, reclaiming the importance of informal politics. Here we show how women's involvement in informal politics is often undervalued, despite their transformatory role in many political organisations. The section also shows how the boundary between formal and informal politics is more blurred than is frequently assumed, and focuses particularly on the importance of political resistance to this blurring. We conclude the section by considering the degree to which gender is (and can be) a mobilising category for political action.

ACTIVITY

Look at a cross-section of newspapers covering the main news items of the week.
- What boundaries are referred to in these current news items?
- What investments are made to maintain these boundaries? Who invests in them? How are they policed?
- Are these boundaries less fixed than they used to be? Do they appear to be breaking down? If so, why do you think so? If not, why do you think not?

- Which people (and groups of people) are likely to be aware of potential break-downs in these boundaries? Why?
- Which people (and groups of people) are not so likely to be aware of these potential breakdowns? Why?
- Try to think of contemporary and/or historical realignments of boundaries which feminist groups have opposed or would be likely to oppose. How can feminist groups resist these realignments in practical ways?

SUMMARY

- Binary categories often structure the ways we analyse space and place.
- Binary categories are invested with power.
- Binary categories often imply a gendered hierarchy.
- Binary categories and the boundaries which divide them are not fixed, although they are frequently presented as if they are.

5.2   Reclaiming the undervalued invisible and blurring the boundaries: work and home; home and work

Focusing on the work/home distinction, this section shows how feminist geographers have reclaimed both undervalued and invisible sides of this dichotomy. Here we show how a research focus on paid work often obscures and undervalues women's participation in the labour force. Moreover, the section shows too how women's activities in the home can be recovered from their usual invisibility in Geography. The section illustrates not only how reproductive work itself has been undervalued but how production is usually dependent on, not separate from, women's reproductive work. In so doing, it also demonstrates that the distinction between home and work is often blurred and indistinct.

Reclaiming work

As we discussed in Section 3.2, the concept of work as applied in both the 'First World' and the 'Third World' is something which is conventionally used in relation to paid work performed by men. Employment classifications were shown to be heavily oriented towards describing and categorising male forms of employment and, consequently, women's waged work was shown to be exceptionally hard to classify using such schemas. As we argued there, the result has been to render invisible women's waged work, and to undervalue this. One response to this on the part of feminist geographers has been to concentrate on the waged work performed by women; to emphasise the significance and the extent of women's paid employment in both the 'First World' and the 'Third World'; to show, in short, that work spaces are inhabited by women, as well as by men, and that these spaces are not just gendered but inscribed with particular visions of gender. As the following two case studies illustrating the work of feminist geographers show, female 'green labour' is particularly important to transnational companies' location not only in the peripheral regions in Britain (see, for example, Lewis, 1984) but in the export processing zones of the newly industrialising countries (Elson and Pearson, 1981), whilst a feature of contemporary labour markets, at least in the UK, is the so-called feminisation of the labour force.

CASE STUDY: Women in the New International Division of Labour (NIDL)

During the 1960s, the world was frequently described and understood as divided into 'core' and 'periphery', with the core countries producing manufactured goods and the periphery, raw materials. Industrialisation appeared to offer an escape for the periphery, and took two major forms. One was industrialisation in countries in the periphery for protected internal markets. Transnational corporations (TNCs) leapt at the opportunity, for instance, to produce goods in Brazil for Brazil, instead of exporting them to Brazil. The other form was for local firms and TNCs to utilise cheap labour in the periphery to produce for the world market. Frobel *et al.* (1980) gave the West a threatening scenario of capital moving jobs from rich countries, attracted by cheap labour, to poorer ones. This dispersal of the manufacturing process has become increasingly sophisticated over time, and, according to Lipietz (1996), is exemplified by the example of Benetton. In these processes, women in particular have come to be involved in production as a primary source of cheap labour. The earliest feminist work on the New International Division of Labour, then, emphasised that the great majority of the new jobs in export-oriented industries in poorer countries (many of them concentrated in Export Processing Zones, or EPZs) were for women and girls, and stressed the ways in which particular feminine (and racialised) attributes were critical to this gendering. An indication of the type of attributes which TNCs found attractive is provided by Diane Elson and Ruth Pearson's quotation from a Malaysian investment brochure:

> Her hands are small and she works fast with extreme care. Who, therefore, could be better qualified by nature and inheritance to contribute to the efficiency of a bench assembly production line than the oriental girl? (Elson and Pearson, 1981: 23)

Feminists, however, have subjected such statements to sustained consideration. They have suggested that girls and women had been educated in manual dexterity and patience, frequently through the tedious tasks associated with domestic labour. They have also highlighted that unmarried women were often preferred, and even sometimes subjected to pregnancy tests. In this way, then, feminists have shown how an image was constructed of women factory workers in poorer countries as young, industrious, naive, passive 'girls' with no experience of union activity: a kind of international teenager (Pearson, 1986).

Reading A highlights the employment conditions of some of the women working in EPZs.

Other feminists, however, have considered this stereotype in other ways. They have explored the geographical variety in the occurrence of women working in global factories and have shown how these processes are concentrated in particular areas and not in others (Pearson, 1986). Another feminist approach has been to focus on different strategies of women's resistance to their conditions of work and the relations of employment (Grace, 1990; Ong, 1987). Therefore, many feminists have challenged the stereotype of poverty-stricken 'Third World' women, ruthlessly exploited by TNCs for low wages under wretched working conditions. Linda Lim (1991), for instance, argues that export factories offer better wages than the alternative jobs available to women in low-income countries. Wages and conditions in TNCs, she maintains, are usually better than in local firms, while some women, as in the best factories in Singapore, are better off than American women in homeworking or sweatshops. The stereotype, says Lim, exists – but in a minority of cases, and certainly women's work in export manufacturing has a complex geography which has generated hot debate.

ACTIVITY

What evidence is there in Reading A and Reading B to support Lim's argument that women in EPZs are not always victims?

Within this debate there is one large puzzle: why so much emphasis on women in export manufacturing? In the mid-1980s there were several hundred million women in paid work in poorer countries. Of these perhaps 1.5 million people worked in export factories, half a million of them for TNCs. Yet this small group captured most of the attention of feminist scholars and organisations. For Lim (1991) this extraordinary distortion derives from the historical coincidence of the growing interest in women's changing roles worldwide with the expansion of export manufacturing in poorer countries. Her point is that the research agenda here was set by Western feminism. From the 1970s, workers and workers' unions in richer countries have felt threatened by the loss of jobs, both to more advanced technologies at home, and to cheaper labour overseas. TNCs in particular were blamed, although most employers in Taiwan, South Korea or Hong Kong are local. Some Western feminists saw the low wages and poor conditions overseas as exploitation which was costing Western jobs. Under this view, TNCs were pilloried as the agents of these changes and the significance of low-paid export manufacturing to job losses in richer countries was greatly exaggerated. In contrast, far less attention has been paid to gender in domestic manufacturing outside the West, although it is quite as exploitative and oppressive, and often also directly or indirectly for TNCs: in Calcutta, for instance, families control the sending of daughters into work for wages which go straight into the family budget, and collude with employers in policing their daughters, so that the work, far from enabling choice and autonomy, reinforces dependence and subordination (Standing, 1991).

CASE STUDY: **The feminisation of the UK labour market in the 1980s and 1990s**

De-industrialisation, or the decline in employment in manufacturing, in the UK has tended to mean the loss of jobs previously defined as male, while the growth of the service sector has been associated primarily with the increasing employment of women (McDowell, 1989). However, many of these new jobs have not only been lower paid than the (male) jobs lost from manufacturing, but they are increasingly part time, short term and/or temporary. Some geographers have suggested that this change represents a fundamental shift in the UK, away from rigid labour market structures regulated by the state and trade unions, and towards a new, flexible labour market – one in which women, with their long tradition of part-time employment, are more likely to find paid work than men (Allen, 1988).

Other feminist geographers have argued strongly that the move towards flexibility in the UK labour market is not necessarily beneficial to women (McDowell, 1992). Far from enabling women to become economically independent from men, the flexible nature of paid work in the UK in the 1980s created a new dependency, especially for single women and those with children. Low pay and part-time work do not provide a living wage, so unless women have a partner who is also in paid work, they end up increasingly dependent on the state, in the form of income support, housing benefit and child benefit. This tendency is frequently referred to as the 'feminisation of poverty'.

Linda McDowell argues further that this era of increased flexibility and apparent equal opportunities for women has created a new division between working women in the labour market. Whilst the growth of service sector employment has primarily resulted in low-paid, part-time and casual work for a majority of working-class women, equal opportunities and the growth of financial and business services has enabled a small, elite group of highly educated women to do very well and compete successfully with men in the higher echelons of the labour market (McDowell and Court, 1994). Other studies, such as Geraldine Pratt and Susan Hanson's work in the USA, support this view (Pratt and Hanson, 1995). However, they also suggest that such women can only succeed in this sector of the labour market by becoming full-time career women, or by delaying parenthood until they are sufficiently high up the ladder to afford full-time nannies (see also Gregson and Lowe, 1994).

In focusing on women's waged work, studies such as the above reclaim the work side of the work/home dichotomy: they bring into bounds spaces which in much of human geography are portrayed as exclusively male worlds, and insist that there is a geography of women to be recovered. They also simultaneously contest the way in which such research frequently represents women's waged work as unimportant, as out of bounds. These studies, however, concentrate for the most part on reclaiming the world of women's employment within the formal economy. In contrast, other feminists, particularly those working in the 'Third World', have concentrated on reclaiming women's role within informal sector employment. An illustration of what is meant by these informal sector jobs is given in the following case study from Latin America.

CASE STUDY: Vendors and domestic workers in Latin America

According to Chaney and García Castro (1989), domestic workers account for at least 20% of the female paid labour force in Latin America and the

Caribbean. Women also occupy a disproportionate number of poor-quality vending jobs (Babb, 1990; Scott, 1991). Empirical studies have revealed that discriminatory mechanisms are at work producing these labour market patterns. For example, women's vending jobs often reflect gender stereotyping, as they produce and sell perishable foodstuffs or traditional artisan products (FLACSO, 1992). In the case of domestic servants, many are young female migrants from rural areas who come to the city to work and study. In Peru, for example, nearly half the female migrants to Lima are aged between 14 and 16 (Ennew, 1993). When they arrive in the city, however, they are faced with long hours of work, which make study difficult to continue. These factors, together with the fact that domestics and vendors are rarely unionised, seldom receive fixed wages, and in general terms become socialised into private, home spaces (Radcliffe, 1990), mean that informal work, especially domestic work carried out on a live-in basis, rarely gives women power. It often only serves to reinforce women's domestic roles as they continue to do what they have always done: work in the home space for little reward or recognition (Chaney and Bunster, 1989). Upward mobility from vending is equally as restricted because women are frequently prohibited from increasing the scale of their operations or from expanding their ventures into non-stereotyped activities due to poor access to credit facilities and discriminatory practices by male retailers, as well as by state and municipal licensing offices (Berger and Buvinic, 1989; Chaney and Bunster, 1989).

Much as with feminist work on the NIDL, research has also emphasised the extent to which women working in informal sector jobs such as domestic work and vending are not always victims.

ACTIVITY

What issues are raised in Reading C which support the argument that women are more than just victims in domestic work in South Africa?

Revaluing home

As well as reclaiming the work side of the home/work dichotomy, many feminist geographers have set out to reclaim women's activities within the home; they have therefore attempted to recover the undervalued and invisible side of the home/work binary. Their aim, in short, has been to bring home space into bounds.

In their work on the middle classes in Britain in the 1990s, for example, Nicky Gregson and Michelle Lowe extend the concept of spatiality into that home space so seldom considered within human geography. They show how dominant ideas about childcare in Britain are both critical to producing home-based geographies of childcare and central in reconstituting home as 'home' (Gregson and Lowe, 1995). They maintain that situating childcare in the home for the middle classes in Britain both reaffirms and reinscribes certain of the meanings of 'home': in short, it helps make home space, 'home'. Another example of research by feminist geographers which reclaims home as a space worthy of geographical analysis is Lynda Johnson and Gill Valentine's article 'Wherever I lay my girlfriend that's my home' (1995). This piece of work looks at how lesbian women create and manage home space. Home here is shown to be inscribed with a whole barrage of meanings, many of them synonymous with the heterosexual family. Johnson and Valentine demonstrate in their article how heterosexual surveillance permeates not just the parental home but lesbian women's own homes too, as well as how lesbian women resist these dominant meanings. Reading D gives an indication of some of their arguments. This essay is a good example of feminist geographers considering the spatiality of gender as it intersects with a range of other social relations: capitalism, but also relations of sexuality in particular.

ACTIVITY

Feminist geographers' work shows home to be invested with multiple meanings. Spend some time thinking about what home means to you. Maybe home is a place of security, safety and belonging for you. Maybe it isn't, or maybe at some point in the past it hasn't been so. Either way, spend some time considering the way(s) in which you perform your identities in different homes. How does this performance vary? For example, how do you vary your performance between your parental home, your own home and other people's homes? And how does this performance connect with dominant ideas about gender (and about sexuality)?

Blurring boundaries

Beyond the activity of reclaiming, many feminist geographers have been concerned to demonstrate the extent to which the boundaries drawn between dichotomous categories such as home/work, and the spaces which they occupy, are more blurred than they might appear to be. The following case study and activity provide an introduction to what we mean here.

CASE STUDY: Teodora tells of her life in Lima

After we were married we went back to Lima to live for a while...I had two children there. We lived there for four years. My husband worked in a shoe maker's shop. He used to do very good work. Really beautiful it was...all for export. But we didn't have much money for him to establish himself.

I really enjoyed living in Lima. For me...I wasn't stuck in the house. I always enjoyed working, or on other days, I would meet people in the street. Because I had worked there before, I knew a few señoras who were very good to me. I would always go to them when I had no money.

And I said, 'Señora, please – do you have any clothes that need washing?' And they said through the intercom, 'Who am I speaking to?' 'To Teodora' I said. 'What a surprise...come in...where have you been...how are you?' they would all say. I would tell them what I was doing, and about all my problems. And they would say, 'Of course you can do some washing...today and tomorrow'. 'But señora', I said, 'I don't even have enough money for the bus'. 'Don't worry', she said, 'I will give you money for fares, and something more too'. And she did give me the money, and I spent the day happily washing. And the other muchachas who worked there all the time gave me lunch. And then, at the end of the day, I went home with my money – and I could buy things for the children, fruit, milk...

I was going to start sewing, so I could make clothes for the children and earn some money that way too. I was just about to pay the tuition fee – just ready – and my husband got sick, and then we just didn't have enough money. We couldn't manage with him being sick if I went to school. There was no one to look after the children, and no one to cook for my husband. So even though he got better, it looked like it wasn't going to be possible.

I would have liked to have stayed there, but we couldn't. At least my children are getting an education here... I will do whatever I have to, whether it means washing clothes or working as a maid...whatever is necessary for them to study. (Bronstein, A., 1981, *The Triple Struggle: Latin American Peasant Women*. WDW Campaigns Ltd in association with War on Want)

ACTIVITY

Using the following list, classify the activities and social relations outlined in the above extract: productive, reproductive, public, private, global, local, employer, employee.

- What problems are associated with these types of classifications?
- Identify the different roles that the people in the extract have.
- What evidence is there in the extract to suggest that domestic work brings independence?
- What evidence is there to suggest that it does not?

Reconceptualising the links between home and work
A further example of a study which examines the connections between home and work, and hence the blurring of the boundaries between home and work, is included as Reading A in Chapter 3, in which Susan Hanson and Geraldine Pratt discuss the inseparability of home and work. These two authors have produced a number of articles – all drawing on empirical work conducted in Worcester, Massachusetts – which illustrate this point still further (Pratt and Hanson, 1991, 1993, 1994, 1995). And further empirical work by feminist geographers has demonstrated this point in a different way, by showing how women often lead 'double days' when they take up paid work, because they carry on with domestic work too: in other words, even when they participate in paid work, women continue to be assigned the domestic role of reproducing the family through domestic work. This is not only the case in the 'First World'. Indeed, as the following case study illustrates, further instances of this double burden can be seen in the case of women in the so-called former 'Second World'.

CASE STUDY: Women and work in Eastern Europe

Judging by certain statistics, it could appear that in much of Eastern Europe and the former Soviet Union women achieved considerable levels of equality with men; a state-sponsored social system appeared to allow women to combine work and family. Indeed, in the former GDR for example, 49% of the labour force was female, with 90% having children at some stage in their lives (Kolinsky, 1993). However, state socialism was a patriarchal ideology. Although men and women were declared equal in state constitutions, women's 'emancipation' was based on the definition of women as both workers and mothers. Furthermore, the range of social benefits was designed to promote a gendered division of domestic and child-rearing work, one in which such tasks were located in the female domain. This double burden was extended with exhortations to women to engage in social and political work, and to gain further qualifications (Corrin, 1992; Tóth,1992). However, it would be wrong to argue that women only participated in the labour force because the state demanded so. Rather, women in the GDR gave family and work equal priority, but found difficulties in balancing both (Bütow *et al.*, 1992). A description of a GDR 'superwoman's' burden illustrates the issues:

> Women are the stronger sex. What man would have managed to be employed full time, to get himself qualified, to bring children into the world and to stand by them, to be active in society and to run a household 'on the side'? To work longer for no pay if it was needed, to look after others when ill, to deal with the school, to stand in queues for food, and all that without complaint? (Rohnstock, 1991: 200)

Waged domestic labour in Britain
Another example of research which looks at the connections between domestic work and paid employment, and hence the blurrings of the boundaries between home space and work space, is Nicky Gregson and Michelle Lowe's study of the employment of waged domestic labour in Britain (Gregson and Lowe, 1994). Their research charts the increase in employment in various forms of paid domestic workers in Britain through the 1980s, notably in nannies and cleaners, and connects this with the expansion in women's employment within full-time, career-based occupations through the same period (and see too the case study above on the feminisation of the labour market in the UK). Middle-class households' solution to the problem of doing domestic labour, then, is resolved in these circumstances by transferring domestic tasks onto another woman, or even a handful of other women – typically of a different class and age to the woman in the employing household. And Gregson and Lowe's study provides considerable detail on the nature of this type of home-working.

This study is an example of feminist geographers problematising the distinction between home and work spaces. However, in order to achieve this problematisation, Gregson and Lowe focus on quite stark differences between women; indeed, the differences between the relatively wealthy women who can afford to employ nannies and cooks and housekeepers, and the nannies and cooks and housekeepers themselves, seem so severe as to question whether the category of 'women' suggests more commonalities between them than in fact exist. This possible displacing of gender as the central analytical category, then, has implications for the distinction between home and work. For the disruption of the conventional gendering of the home throws into question just how this space is a 'home'; the waged labour of some women makes it a 'home' for their employers, but there is little possibility that, as a place of waged work, it can feel like home for them. This suggests that home spaces may be contradictory and shifting, remade and displaced; the complexities of their gendering disrupt conventional meanings associated with home space, requiring us to break with straightforward identifications between home and feminised reproductive labour (see Section 5.1) and to constuct in their place more complex theorisations of the spatiality, and the gendering, of home.

ACTIVITY

Spend some time thinking about the employment of women as nannies, cleaners, etc., within the home. Why do you think this type of work is gendered as female work? What do you think about this type of employment: do you think, for example, that it is a progressive development, in the sense that it could be argued to provide women with employment; do you condone this development, in the sense that current arrangements in Britain – particularly with respect to childcare provision for parents in full-time employment – provide few alternatives; or do you think that it reinforces existing gender inequalities with respect to domestic work? Why do you think that these households employ waged domestic labour in place of getting the men in their households to do more in the way of domestic work?

If Gregson and Lowe's work suggests that the gendering of home space can be highly complex and even contradictory, other feminist geographers are beginning to suggest that when the home is a waged workplace it may also be a site where new femininities can be negotiated. This can be seen clearly in the following case study from Mexico.

CASE STUDY: **The Mexican sweater trade**

Fiona Wilson's work on the Mexican sweater industry (Wilson, 1993) raises some interesting issues. She suggests that in the context of small-scale domestic production (homeworking) women can be more than victims; in short that they contest established gender identities 'from within'. Her work illustrates how two important themes come together to create an environment in which women can contest dominant femininities. These two themes are the image of the household as a protected domestic space and the physical concealment of deregulated economic activities. These factors work together to create a space where young women can have new experiences that encourage them to challenge established views concerning their appropriate behaviour and activities, while at the same time providing the veneer of a respectable domestic environment rather than a factory for this to occur in. These circumstances set up a series of contradictions which may be used to renegotiate gender identitities and, in particular, femininities. Reading E outlines some of Wilson's arguments.

What Wilson's work begins to do is to point to the importance of waged work spaces – whether these are located in or out of the home – as spaces which create opportunities for gender identities to be renegotiated, perhaps even to be reconstituted, and in which new forms of femininity might emerge. The following two case studies provide examples where we would like you to think a bit more about workspaces as potential space for change.

CASE STUDY: An Italian clothing factory

'Traditionally women have always worked in the textile industry and in the production of intimate apparel. But how much being endowed with a pair of breasts or a rear end counts for good production is subject to verification. Does being a consumer of what they produce make the employees more expert than most in matters of panties and bras? Some managers think so. A work of true excellence? Only in the narrowest sense: "The precision work is the most boring: tiny stitches" states Muccia dispassionately, "sewing a bra is like sewing a jacket". The pleasure of creating a certain piece of lingerie? "There is no satisfaction, working on an assembly line you think only of numbers. Also you don't have the finished product in your hand". So it is the buzz of the enormous bobbins of thread, needles on the machines trimming the edges, feet pushing the pedals, and arms threading the hooks, sewing the shoulder strap, adjusting the lace, packing, wrapping. Every day they send out almost 7000 pieces in 15 styles. "All fresh" assures Valentina, "just like a pastry shop."

'Valentina is an American-style "self made woman", she has the personality of a housemother. The rise from machinist to director came about, recalls Valentina, thanks to her "monkeyish" curiosity. Today, Valentina directs production with a chaotic efficiency, "when I put passion into my work, it is certain to be passed on to the others" is her management style.

'The others are the 350 workers employed as labourers. "Their problems are my problems", says Valentina. She speaks of her "young women" as if they are children and a part of her. "I am a worker as they are workers. I am they and they are me". A symbiosis not shared by the workers themselves. "The factory runs on sympathy and antipathy" Anna explains, "there is good and bad. If you don't create problems and stay at your post your career is secure".' (*Connexions*, 44 (1994))

CASE STUDY: Peru – state-backed Temporary Employment Programme

Between 1985 and 1990, 80% of the national workforce in this programme were women, who worked under renewable three-month contracts. The following quotations are taken from Laurie (1995).

'The work was hard, hard, hard, too hard in fact. It was totally killing, digging ditches and carrying bricks. We even made bricks to build a containing wall around the stadium and we also made the floors.'

'We climbed up the hills and excavated the rocks, then the other group of women made like a human chain and we passed the rocks along and placed them on the road: rocks on top of rocks. Then, on top of the rocks we put small stones so that lorries carrying water could get up to the shacks.'

'For us there was no such thing as hard work because in the campo (countryside) we practically climbed up with crowbars and dug out the rocks. We carried some enormous rocks. So we worked harder than a man.'

'In the government programme I learnt all about responsibility. For example, to work in a factory you have to work to a fixed schedule, with a starting and finishing hour. There (in the programme) we had to do the same. We had an hour when we had to start, a time for rests and an hour to leave. More than anything it was about co-ordinating things with other people so that we could agree on "you do that and I'll do this", that's the same as it is in a factory – you do that, I do this – the only difference is that we work on the land. The work in the programme was about working with rocks whereas in the factory you work with processed materials.'

ACTIVITY

The case study on the Italian clothing factory shows how workspaces can both reinscribe dominant visions of femininity (here women literally create garments which define femininity) and be a site where such ideas are resisted (sewing lingerie is only numbers).

- Looking at the Peruvian case study, how do the spaces of the Temporary Employment Programme enable women to think differently about their gender identities and femininity?
- In both case studies what divisions of labour exist? What ideas shape the worker–tools–technology relationship (think back here to the NIDL case study)? How does the nature of the finished product appear to influence ideas about appropriate work for particular people?
- What evidence is there in the Peruvian extract to show how workspaces can be simultaneously reconstitutive of dominant ideas about femininity and challenging of these?

SUMMARY

- Feminists have reclaimed women's contribution to both formal and informal work.
- Feminists have stressed the significance of home, the undervalued side of the home–work dichotomy, and the different meanings which are inscribed in this site.
- Feminists have undermined the home–work binary by showing that the boundaries drawn between home and work are blurred.
- Feminists have shown how workspaces are frequently spaces in which gender identities are negotiated, resisted and changed; where new and old, dominant and resistant, forms of femininity may be found alongside one another.

5.3   The politics of boundaries: reclaiming and blurring the formal and the informal

A second major area of empirical work in feminist geography focuses explicitly on the division which is made between formal and informal politics, and it is this work which we explore in this section. As we show, whilst some feminist research has used the formal/informal dichotomy to analyse political activity and to reclaim invisible activities, other research has questioned how accurate such categories are. In what follows, therefore, we address the ways in which feminist geographers have looked at the formal/informal divide, focusing particularly on the ways in which both the formal and informal sides of this dichotomy have been reclaimed, and on the blurring of the formal/informal political boundary.

Gender-blind formal politics

'More than any other kind of human activity, politics has historically borne an explicitly masculine identity' (Brown, 1988).

Few would argue with the assertion that masculine assumptions and a masculine identity pervade formal politics, defined conventionally, for example, as elections, parties, parliaments and unions; although the character of formal politics is diverse, everywhere this is largely men's territory. Thus, and despite a few significant exceptions, there are very few women in leadership positions in formal politics, and women remain dramatically under-represented in formal political institutions. This lack of access to formal power structures and decision-making bodies at international and national levels helps explain gender inequities around the world (Barker, 1994; Elson, 1991; Ford *et al.*, 1994). Moreover, such a situation results in 'women's issues' – for example, reproductive rights, rape, domestic violence – being considered as separate (and inferior) political issues; issues which can be relegated to the personal arena, where they may be seen as someone's own private business or as issues of personal choice, in short as out of bounds of formal politics.

Multiple factors help explain the invisibility of women in formal politics, and accounting for this invisibility has been one way in which feminists have attempted to reclaim the formal side of the formal/informal political dichotomy. Included amongst these reasons is the way in which ideologies which define femininity and masculinity work to constrain women's access to formal political power. Thus, the traits which are frequently associated with success in the formal political arena – aggression, ambition, authority, for example – are labelled as positive male attributes, but as unfeminine, whilst femininity defined in terms of motherhood is often argued to be incompatible with formal politics as currently organised. Such arguments help to explain women's poor performance in the numbers game of formal politics, and this in turn is why some organisations have opted for policies of proportional representation by gender and/or quota selection systems.

ACTIVITY

In Britain, the Labour Party recently attempted to increase the number of women among its potential Members of Parliament by designating women-only short lists for candidate selection. When this was first proposed, some men (and some women) within the party argued that these lists represented an act of sex discrimination. After an industrial tribunal ruling in February 1996 declared the policy in breach of sex discrimination laws, the policy was dropped by the Labour Party amidst much controversy and debate.

- What do you think the issues are in such debates?
- Make a case for women-only short lists.
- Make a case against.
- Suggest at least three other ways in which women's involvement in party politics could be increased.
- Using newspaper archives, establish how well the points you've made coincide with those made at the time of the debate.

A further way in which feminists have reclaimed the formal side of the formal/informal political dichotomy has been to show that there are other, less obvious, ways in which formal political discourses are gendered. One example of this is the way in which national identities are gendered by the state.

National identities are based on identification with a given territory, which in many cases is also defined as the state. How states act in domestic and international arenas is the conventional stuff of political analysis in geography. Thus, conventional political geography includes analyses of the ideologies, apparatuses and capacities of the state (Taylor, 1994). That states are structured by 'race', class and ethnic alliances and antagonisms has long been recognised, but most conceptualisations still fail to appreciate the gendered character of discourses on nationality and the state. Nonetheless, what Anderson (1983) refers to as 'imagined communities' are always gendered communities (as Chapter Six also points out). Nation states remain powerful agents in forging political, economic, social, gender and sexual identities (see, for instance, the recent discussions of citizenship in the journal *Political Geography*, 1995), and the development of nationalism remains one of 'the most politically salient, socially constructed form(s) of cultural/regional identification extant in the world today' (Reynolds, 1994: 234). These powerful discourses are gendered in complex ways. For example, Enloe (1983), examining the role of the military in shaping ideas about masculinity and femininity, argues that militaries not only need men to act as 'men', by being willing to kill and die on behalf of the state as proof of their manhood, but need women to behave as 'women' (see also Elshtain, 1987; Elshtain and Tobias, 1990). As 'women' they are constructed as mothers of (male) citizens. Such ideas are exemplified by the following case study of the role of women in Nazi ideology.

CASE STUDY: Women and Nazi ideology

The position allocated to women in the ideology of the Third Reich is encapsulated in the phrase '*Kinder, Kirche, Küche*' (children, church, kitchen). In other words, women were defined in terms of reproductive activities, as wives and mothers, which in turn were seen as critical to the creation of the nation. These ideas were closely interwoven with the Nazi ideology of race, which stressed the organic connection between blood (the purity of which guaranteed the purity and the strength of the Aryan 'race') and soil (the land which anchored the people to their *Lebensraum*, or living space). Ernst Loewy (1966) argues that, in this ideology, women became simply machines for giving birth, since the idealised link between blood and land ('race' and nation) effectively equated 'mother' with 'earth' (see also Section 6.4). He writes further that 'this is a motif which is as old as literature itself, but what is new is the way it effectively dehumanises the woman to make her simply a female.... The seeming contradiction between the aestheticised picture of the chaste virgin (who has to keep herself pure under the German race laws to protect the nation) and the earthy bearer of children is only apparent. Both are founded on the same values: it is the view of a man's world which recognises women only as servile beings'. This view was reinforced in the ideal of the 'blonde, blue-eyed, sturdy and broad' German girl, as presented in the girls' branch of the Hitler Youth (the Association for German Girls – BdM), who 'had to reject as indecent any hint of charms or coquettishness. The erotic had no place in a love which solely served a purpose; it had to be subsumed in the act of procreation which served only the purpose of delivering to the state children and cannon fodder' (Loewy, E., 1966 *Literatur unterm Hakenkreuz*, Europäische Verlagsanstalt, Frankfurt: 114–115).

ACTIVITY

Think of two examples of conflicts or wars, one historical and the other contemporary or recent.

- What roles have been ascribed to women and men in these conflicts?
- What qualities is each gender praised for?
- What factors help make these images seem monolithic?
- Whose interests become invested in these images? (In the examples you have chosen think of institutions as well as people)
- How are these images resisted and challenged?

In and out of bounds
Another way in which we can understand something of the gendering of formal politics, and thereby understand its engendering, is by focusing on the way in which gendered states place some people 'in' and others 'outside of' bounds. One of the starkest, and perhaps the best known, instance of this process of defining people as either in or out of bounds is the recent policy of apartheid in South Africa: apartheid not only made some people (black people) dependent on a pass system, but also excluded women from gaining access to certain spaces, such as all-male hostels which housed migrant black men working on short-term permits. As well as apartheid, though, migration provides a particularly good example of the gendered state. This can be seen clearly in the following case study of the female refugee.

CASE STUDY: The female refugee – the ultimate victim?

Current debates suggest that the hegemony of the state is seriously challenged in an increasingly interdependent, globalised world economy (Anderson *et al.*, 1995). In these debates we are invited to imagine alternative geographies of the world, where the status of the nation state is usurped as the most important organising unit in the world economy. Tensions from within and beyond state boundaries are certainly testing the institution of the state and its capacity to survive, yet whilst capital flows are relatively free from the constraints posed by national boundaries, labour (people) continues to be constrained by immigration laws based on the nation state unit.

Many studies have sought to show how immigration policies affect patterns of migration, for example through looking at recruitment and asylum policies or at work permits for skilled migrants. However, these policies have a strong gender component. Labour recruitment policies, for instance, often seek women for specifically 'female' jobs, while legislation on recruiting overseas workers as domestic labour largely affects women. Janet Momsen (1992), for instance, shows how migration from the Caribbean is strongly influenced by state policies, such as the Canadian Domestic Worker programme which encouraged women to migrate, while changes in US immigration policy towards more skilled migration mean that Caribbean women who are teachers or nurses can gain access to the US more easily. Historically, though, women have not been the primary migrants. Ideologies of the roles of men and women generally see the male 'breadwinner' migrating, while the woman either remains at home or follows on after. However, in the contemporary period it is increasingly the case that migration between 'Third' and 'First' World countries involves women, and one particular form of contemporary migration which clearly illustrates the gendered nature of the state is that of the refugee.

The official United Nations Convention Relating to the Status of Refugees defines such a person as someone who

> Owing to well-founded fear of being persecuted for reasons of race, religion, nationality, membership of a particular social group or political opinion is outside the country of his nationality and is unable or, owing to such fear, is unwilling to avail himself of the protection of that country; or who, not having a nationality and being outside the country of his former habitual residence as a result of such events, is unable or, owing to such fear is unwilling to return to it.
> (quoted in Forbes-Martin, 1991)

All refugees inhabit a space of extreme marginalisation and vulnerability by having moved outside the established pattern of states and citizenship, and female refugees share certain aspects of being 'out of bounds' with male refugees: loss of state legitimisation by crossing physical boundaries, and losing a sense of belonging. In this situation there is a need for them to reconstruct their own boundaries and identities in a 'foreign' place, while other people are imposing *their* boundaries (and identities) on them, for example through actions which might range from holding pens for signing forms to keeping people policed in camps. However, there are also experiences specific to the gender of the refugee.

In 1991 there were approximately 20 million people recognised as refugees by the UN High Commission for Refugees (UNHCR). Even greater numbers of people form the category of displaced people. These are people who are internally displaced and although equally disadvantaged as refugees, they are not classified as such. About 80% of these uprooted people are women and children, many of whom live in women-headed households (Forbes-Martin, 1991).

ACTIVITY

Study Figure 5.1 and answer the following questions:

- What experiences do displaced people have in common with official international refugees?
- What differences are there between the experiences of these two groups?
- What are the similarities and differences in the opportunities and problems which face women and men when they become refugees?
- Choose one country shown in the figure and find out more information about the causes and gendered effects of displacement in that country.

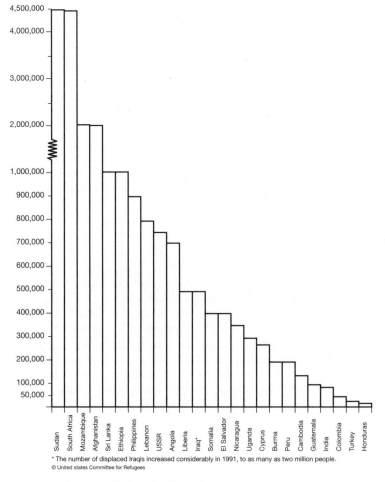

* The number of displaced Iraqis increased considerably in 1991, to as many as two million people.
© United states Committee for Refugees

Figure 5.1   Internally displaced civilian refugees

One of the things which you might have established from the above activity is that when female dissidents are imprisoned within their country of origin, the tactics employed by their guards and interrogators are different from those suffered by male internees. Agger (1992) describes various of the methods used throughout much of the world to control and punish women who are seen to constitute a threat to the state, and sexuality plays an important role in this. Once women move outside the bounds of the state, particularly by becoming refugees, rape and prostitution are again seen as 'natural'. Indeed, the rape of women in war zones is frequently perceived as a legitimate tactic, as a means of emasculating and demoralising opponents through the identification of woman with nation.

Informal politics: the domain of women, new spaces for women?

As well as reclaiming the formal side of the formal/informal political dichotomy, feminists have examined the less valued and frequently invisible side of this dichotomy, that is the spaces of informal politics. In other words, they have brought such activity and such spaces into bounds. Thus, women have been shown to be strongly involved in local, community-based activities, campaigning for example around issues to do with housing, education, food, safe play areas for children, and so on and so forth. Moreover, women can be shown to be becoming increasingly involved in global networking activities outside formal political institutions. One example of the latter tendency is illustrated by the Fourth UN Women's Conference, held in China in 1995. Overlapping in terms of both time and participants with the formal UN conference was an informal forum of non-government organisations (NGOs), to which 46,000 women from many countries (and a few men) came to share ideas and information and to meet and discuss many issues. Although the Planning for Action document, to which all countries could commit themselves, was negotiated within the formal conference, there were many spaces at the conference within which women from the NGOs could, and did, directly influence the course of the negotiations. This resulted in some important changes to the final document (Foreign and Commonwealth Office, 1995). This gathering was an example of women asserting their presence on the world stage and making their demands heard as women.

Some of the best illustrations of women's involvement in informal politics focus on women's involvement in popular feminisms. Popular feminism is a term used to refer to a wide variety of activities which involve women acting collectively, usually around a basic need which often reflects what are assumed to be female, reproductive responsibilities (tasks such as food preparation and cooking). These activities create opportunities for women to think differently about themselves and sometimes provide the space for them to create new gender identities. Both of the following case studies illustrate all these features, and are drawn from Latin America. Such activities, however, are not confined to women in countries of the South, and one oft-quoted case from Britain is the role of women's organisations in the miners' strike of 1984/85 (Miller, 1985).

CASE STUDY: **Coming of age in Conchali**

' "When I was a girl I never went out and did anything. But now, at 63, I've changed completely. If anyone says let's go somewhere, I go. I'm really enjoying myself." Pilar is part of a network of older women in the marginal borough of Conchali to the north of Santiago in Chile. This network grew from a nucleus of 10 older women who decided they wanted to do something for themselves and their community. They were trained as promotoras, or community workers, by the Instituto de la Mujer, a women's group in the capital and have been busy organising the community via older women's groups ever since.

'Lidia is in her sixties and is one of the original 10. She began as a promotora by carrying out a survey to find other women of her age. Lidia says: "We started just by talking through our problems and I began to run some relaxation sessions from what I'd been taught. For a while that was all we did. They said they wanted to do some more, so we got started."

'Lidia has been helping older women to come to terms with approaching old age, by holding discussions on a whole range of subjects, including the role of older people, health care, basic rights and sexuality. The women are then equipped to work in the community with others – both young and old.

'Laura was one of the first promotoras to be trained. She found that her skills as a gardener and farmer were much in demand. So she started small-scale farming sessions, creating roof top gardens in the densely populated borough. Once a week over 20 older women meet and learn about seeds, watering, transplanting and making bedding materials.

'Personal development, alongside community organisation and involvement, has been the key to the mushrooming of the older women's network. Germania speaks for many when she describes her thoughts on her involvement: "I came with the idea of learning and that's what I've done and one of the most important things that has helped me most is having self confidence." Germania is now a representative on an older people's parliament which has been formed in the area.' (*New Internationalist*, February 1995)

ACTIVITY

- This is not a project for all women. Why do you think it specifically targets older women?
- What new roles and experiences is the project giving these older women in Conchali?
- What positive effects is the project having in the women's lives?
- Think about other questions which you would like to ask these women to find out more about the significance of this new space in their lives. Design three questions you would ask of each of the following individuals: Pilar, Lidia, Laura, Germania and the author of the piece?

CASE STUDY: **Argama Mothers' Club**

In 1989, with the help of state organisation, the local women in Argama (southern Andes, Peru) had a building erected for their mothers' club. The location provided a meeting place for the women, where they co-ordinated their cooperative project. They were given seeds by the government and grew vegetables to keep and to sell. They also grew seedlings which they were allowed to take away and cultivate on their own land. Sometimes through this club they received food aid. What follows is the text of an interview with the club's president, Naomi Velasco, which was carried out in December 1993 by Nina Laurie.

NV: Really I formed the mothers' club in order to work in literacy

NL: What year was that?

NV: In 1986. I didn't know how to involve the women. To encourage them to come you have to make a group and so I formed the mothers' club by contacting about 25 women. The women told me that women get together more when there is food and so I started to go to the local government. One señor came and said, 'We are going to give food but only when you work'. So we started to make the people work.

NL: What did they work in?

NV: We prepared the soil to grow potatoes, beans, wheat; we made family gardens, all that, and so the people gathered together more, more than 110 señoras from the 25 that we were, and now in total there are 210 señoras.

NL: Did the government give you any money?

NV: Yes. When we finished the building they paid for the teacher, the señorita, for the crêche/ nursery school.

NL: Who worked in building the centre?

NV: They were employed by contract. They gave us a builder, a Sr Pablo Rincon.

NL: You, the señoras, you didn't work on this building?

NV: In food preparation, nothing else. There was another one, a carpenter who was in charge of the walls, another one in charge of the floors. It was paid by contract and the community gave its labour, it always gave 10 people, something like that.

NL: And when the community gave the labour, what were your tasks as señoras? What did you do?

NV: Food, yes food.

NL: You never worked building walls and things like that?

NV: No. Food, nothing more.

NL: You all put your bit in for the food? You provided it yourselves?

NV: Yes, from us and when they inaugurated the building each woman came along with her little guinea pig. I told everyone to bring their potatoes, their vegetables and their guinea pigs to cook. [In Andean Peru, guinea pigs are reared for food.]

NL: Señora, this work with the club and all that, what does it signify for you as a woman and as a member of this community?

NV: Well, all I say is this is to open the eyes of the señoras because there are señoras who in reality cook the same thing, for example they cook wheat, wheat alone. I have taught them to cook things mixed up, things from this area. In one way it has helped me a lot too because instead of going up I was getting dragged down with all my problems. I was depressed with my problems because I'm not from this place. I'm from Lima, my husband is from Argama.

NL: And what did your husband think about what you were doing?

NV: 'You're just getting involved because you want to. You're just cocky' he said to me. 'What are you doing getting involved with all those women.' But it's not like that. When you get to know about the problems of the women you can help them, you feel then that you have something, that you give something to the people, that you help. In contrast, before, when I was in the house, I was depressed.

NL: Has your husband changed his attitude?

NV: Yes. He's got used to it. He saw that sometimes I brought the people here and he saw that it was as if I were a person that had value. Then he said that his wife was another person, like a leader, I had that respect.

NL: What did he say about the money you earnt?

NV: It was a small amount but it was a help because here in the campo [countryside] we don't have much of that sort of life. There isn't money, there isn't work. So when I came into the house I arrived with something. He was pleased and with that he then helped me more. We the women did an event to get money to buy pots, pans and plates, cups etc. We had a football game – just between the women. I made them play football! It was to encourage them so that they would forget the problems in their house. We all played football. It was funny, let me tell you, because the señoras forgot everything. They left their children on the ground on one side and one señora looked after them, and then they changed into goal keepers, centre forwards – these were the same women from the mothers' club! It was very funny, they were all playing in polleras [homespun skirts].

ACTIVITY

- Given the fact that Naomi's original idea for the club was literacy training, why do you think she was encouraged to switch to food production?
- Do you think this strategy was successful?
- Why might it be difficult for Naomi to work in this group?
- How did she deal with such difficulties?

Now, thinking across the two case studies (Conchali and Argama), try to answer the following questions:

- What seemingly 'traditional' gender roles are expressed in these two extracts?
- What seemingly 'non-traditional' gender roles are expressed?
- What do these extracts suggest about changing gender roles in the context of popular feminist activities?
- What different impressions do the two extracts give you about the potential of popular feminist spaces for generating new femininities?

As both the above case studies show, the spaces provided and created through popular feminist activities frequently create opportunities: much as we saw with the examples of Mexican workshops, the Italian clothing industry and the Peruvian Temporary Employment Programme in the previous section, the spaces of informal politics provide opportunities for gender identities to be negotiated and rethought. New constructions of femininity can be forged, as for example in the Argama mothers' club, and they can be forged in the context of political activities which focus on both 'traditional' and 'non-traditional' female activities.

Blurring boundaries

Just as feminist geographers have been concerned to demonstrate the blurring of the boundaries drawn between home space and workspaces, so feminist geographers working on political activities and political spaces have challenged the distinctions drawn between formal and informal politics. One example of a study which does this is included as Reading B in Chapter Three (see also Rose, 1988). This is Sue Brownhill and Susan Halford's comparative study of the actions of women involved in two spheres which would conventionally be labelled 'formal' and 'informal' politics – women's committees within local government in Britain and a 'community'-based protest group campaigning against developments in London's Docklands respectively. Their research, however, shows women's political activities to be very fluid: strategies from the formal political arena are deployed by the supposed 'informal' political group, whilst the women's committees use tactics from the so-called 'informal' arena. In short, then, their research calls into question the validity of the boundary drawn between formal and informal politics. Moreover, Brownhill and Halford argue that the boundary between formal and informal politics is one which reinscribes old gender divisions. For them, formal politics is all too frequently associated with male forms of political organising, whilst informal politics is assumed to be the domain of women. In subverting these assumptions, their work not only shows the gendering of the formal/informal dichotomy to be false, but reveals how political research has responded to the challenge posed by the women's movement – by recognising women's political activities, but by boxing these off in the (undervalued, feminine) category (and spaces) of informal politics.

ACTIVITY

Read Brownhill and Halford's paper (Chapter Three, Reading B). Then, using either local newspaper sources or personal experience, identify an example of informal politics. To what extent does this example appear to be represented in the local media (press, TV and radio) as women's politics? Think about who is pictured and who acts as the spokesperson for the group as well as about the issue itself. What strategies are this group using? Do they appear to stem from the formal or informal political domain? Spend some time thinking through the formal/informal political dichotomy. How valuable did you find this in shaping your thinking about your chosen example? Do any of your thoughts connect with the arguments made by Brownhill and Halford? Do you think that Brownhill and Halford's arguments apply to your example? Alternatively, do they not appear to do so? What does this suggest to you about the general points which they are making about the gendering of the formal–informal dichotomy?

Whilst feminist research such as that of Brownhill and Halford has shown that the lines drawn between formal and informal politics are frequently blurred, this position still leaves us with the question: to what extent is gender (and should gender be) a mobilising category for political action? In what ways does the category gender have potential and limitations for finding common identities upon which to act politically? One answer to this question is provided by the Bolivian political activist Domitila Barrios de Chungara, who expressed her doubts about alliances of gender at the UN Conference for Women held in Mexico in 1975. She said:

> Those weren't my interests. And for me it was incomprehensible that so much money should be spent to discuss those things in the tribunal...for me it was a really rude shock. We spoke very different languages. (Barrios de Chungara, 1978)

Domitila's personal experiences were grounded in class struggle, and focused specifically on the closure of the Bolivian tin mines. For her, white, Western and middle-class concerns with sexual freedom were unimportant.

It is all too easy to assume, and is often argued, that alliances around gender are somehow inherently progressive. The following two case studies begin to question this assumption. The first focuses on the experiences and actions of women in Eastern Europe, and shows how femininities are being redefined in relation to the restructuring of work, home, the state and the body. These new definitions challenge Western assumptions about what makes for 'progressive' feminist geographies and show the difficulties in finding common alliances around gender. The second case study, this time from South Africa, shows the ways in which other alliances can disrupt projects around gender.

CASE STUDY: Women in the Soviet Union and Eastern Europe

The experiences of women in the former Eastern Bloc countries are now taken to challenge many of the assumptions made from Western viewpoints about home, work, public and private spheres. Women in Eastern Europe now face a disproportionate burden of unemployment. In eastern Germany women make up 55–60% of the unemployed (Bütow et al., 1992). Russian female unemployment rates are twice those for men (Bruno, 1995). Women also fill different positions in the transforming employment structures. Work by Marto Bruno showed that in Moscow, for example, men engage more in entrepreneurial activity while women concentrate on more secure employment in state companies and foreign firms. High female unemployment rates also reflect the extent to which employment in the past varied between men and women. Now, however, many women seem relieved to be able to return home and their collective position in the new labour markets has not caused mass demonstrations.

In the context of the double burden, their current exclusion from the labour force is viewed by many women with ambivalence. Social services have become more difficult to find or are increasingly expensive, forcing more women to take responsibility for the elderly and for children, while unemployment increases the feminisation of poverty (Tóth, 1992). In this situation a cost–benefit analysis of childcare costs and income levels makes staying at home a logical option for many women (Bruno, 1995).

Yet many women also see a return to motherhood and housework as an escape from the double burden. A study of women in Moscow showed that the 'luxury' of being a housewife was a desirable, if difficult to achieve, objective given the dominant gender divisions where women are responsible for the production of the private sphere while men are involved exclusively in its consumption (Bruno, 1995). In the past, family life was the only sphere of relative autonomy from party and state. 'Home' had values associated with refuge and independence from state involvement. However, the sharp falls in fertility rates suggest that rather than returning home to become mothers, many women are simply excluded from the labour force and would work given the opportunity (Dölling, 1991). That these patterns vary shows how economic and social restructuring intersects with cultural values and expectations about the status of household work and the values of paid employment, both of which are linked to specifically gendered positions.

It is not only in the spheres of work and home that Western ideas are challenged in Eastern Europe. The Western feminist motto that the personal is political and the political is personal clashes with the bitter experience of many East European women of the incorporation of personal choices into state policies. In this situation state-sponsored 'emancipation' was equated with gendered divisions of household responsibility combined with paid work. The retreat by some women to a private sphere represents a rejection of this burden and with it political action (Bütow et al., 1992; Tóth, 1992). In the Czech Republic and Hungary there are now strong anti-feminist sentiments (Einhorn, 1991). This is seen particularly in the conservative backlash against abortion rights as they are seen as part of the former system or are attacked by the Catholic Church and by strong nationalist cultures which construct women as the national baby producers (Dölling, 1991).

While some Eastern European feminists share similar formulations of issues with Western feminists, there is a strong 'allergy' to Western feminism even among women activists in many Eastern countries (Funk, 1992; Tóth, 1992). Western women, it is argued, draw on an 'unbroken consciousness' in their aim to create the private as political (Einhorn, 1991). While Eastern women still appreciated the positive values of paid employment (Rohnstock, 1991), now they often feel they have to 'catch up' on Western women's material benefits and 'sophistication' (Funk, 1992) and at the same time feel Western women are patronising them:

> Western feminists have distorted the situation of women in eastern and central Europe in two ways. The first was distorted by envy, expressed in their claim that women in Eastern Europe had 'by some fluke', 'gotten it all'... . Both the

communist state and Western feminists emphasised that these possibilities were not available in the West... . The present situation... is that Western feminists treat Eastern European women with trembling compassion... they wonder how we survive it at all.... I do not want readers either to envy or pity us, but to understand us. (Tóth, 1992: 213–214)

CASE STUDY: South African women in the 'new' South Africa

The post-apartheid South Africa was officially designated the 'non-racial, non-sexist Republic of South Africa'. One-quarter of the country's Members of Parliament are women. The new constitution also sets out the aim of a Gender Commission to implement the 'non-sexist' commitment. Bridget Mabandla, an ANC MP, argues: 'We transformed the parliamentary tradition... and we bamboozled the other parties. Not only with our garb, but they had never seen so many articulate women in Parliament' (Haffajee, 1995).

However, these aims and ideas have been slow to materialise. Much of the strength of the women's movement prior to the new Parliament being established was based on good, widespread grassroots organisation. However, the strength of these non-state organisations was undermined when many of their leaders were either voted into the country's Parliament or were attracted to work in government. 'Women's organisations that had kept the gender flame burning suffered the loss of both key staff and funding, as aid agencies began channelling money straight to the now legitimate Government' (Haffajee, 1995).

Such problems relate to the other alliances in which women were involved. One researcher, Cathy Albertyn of Witwatersrand University, argues that because women in South Africa historically organised through the liberation movements, which have now become political parties, 'gender was subsumed to national liberation because people responded to the more obvious source of oppression, which was race rather than gender' (Haffajee, 1995: 13). Many women are now interested in working for change within these organisations, such as political parties. However, others argue that other alliances must be developed and that the particularly poor position of many women must be raised as a key issue. Black women still make up the majority of the unemployed and poorly paid, make up the majority of the poorest rural households, are bound by 'the yoke of traditional leaders who control their access to resources; and they are bound by discriminatory marital and property laws' (Haffajee, 1995: 12). In addition, violence against women, particularly rape, is an issue around which various organisations have been campaigning for decades.

Some women have argued that it is not enough to rely on the established structures of power (such as the women MPs) to change the situation. They want 'a national independent women's movement which will put the interests of women first, as opposed to the interests of political parties, churches or unions' (Debbie Bonnin, quoted in Haffajee, 1995: 13), and women MPs are beginning to seek to organise on broader lines too.

ACTIVITY

The South African case study demonstrates that there are other sets of interests which cut across alliances around gender. These can simultaneously undermine and strengthen gender alliances. Re-read the case study and make sure that you can identify the ways in which other interests simultaneously undermine and strengthen gender alliances. The Eastern European study emphasises the degree to which home and motherhood are open to very different interpretations to those which have dominated Western feminist thought. Can you see any potential difficulties which there might be in trying to find a common agenda for action between Western and Eastern women?

SUMMARY

- Feminists have reclaimed the formal side of the formal–informal political dichotomy by showing how formal politics is gendered.
- Feminists have reclaimed the informal side of the formal–informal political dichotomy by examining the importance of women's participation within informal politics.
- Feminists have challenged the boundaries drawn between formal and informal politics by showing that the two are strongly connected.
- Feminists have shown how the spaces associated with informal politics are frequently potential spaces; spaces in which gender identities can be negotiated, resisted and changed; where new and old, dominant and resistant, forms of femininity may be found alongside one another.
- Feminists have shown that gender is an important category for political mobilisation.

5.4   Conclusions

In this chapter we have illustrated the diversity of feminist geographers' engagements with space and place. Feminist geography has usually understood space as relational, and has both analysed the gendered production of dichotomous spaces and explored the ways in which such dichotomies can be problematised. The chapter shows that whilst feminist geographers' notions and uses of space and place are often structured by dichotomous thinking, feminist geographers work with and against these dichotomies in various ways. Thus, whilst much research has worked from within dichotomies of home/work and formal/informal politics in order to make important points about the exclusion of women's experiences, other work has pointed to the blurred nature of binary categories. In other words, it is challenging dichotomous thinking, and the boundaries which this sets up and polices.

Given the embeddedness of dichotomies like that between home and work, or the formal and informal sectors of the economy or polity in contemporary societies, it is perhaps not surprising that feminist geographers have paid them so much attention. This attention has certainly been one of feminist geography's major contributions to feminism more generally. However, we would like to end this chapter with the suggestion that perhaps feminist geography needs to push its diverse interests in the complexity of gendered geographies even further. Thus perhaps feminist geographies need to consider starting with non-dichotomous frameworks of analysis. What is needed, then, and something which feminist geographers are only just beginning to think about, is a non-dichotomous way of thinking about space and place. In the conclusion to this volume we introduce one of the provisional ways in which feminist geographers have begun to explore such possibilities.

A second point to emerge strongly from this chapter is that significant silences occur within feminist geographers' research on space and place. Whilst on the one hand feminist geographers have been keen to discredit universalising assumptions about home/work and informal/formal binary categories, on the other hand our efforts have largely (although not exclusively) taken place within an economic paradigm. That is to say that, although research has demonstrated how notions of work should be expanded to include homework, housework, waged domestic labour and so on, much of this research has occurred within a debate around productive and reproductive labour, and is positioned within the spaces of production and reproduction; spaces which are often still divided into public spaces (of work) and private spaces (of home and the family). Not only does such research reinforce other dichotomous categories (notably public/private), but it is clearly linked to the continued influence of socialist feminist agendas in feminist geography (expressed in *Geography and Gender*) and to the colonial origins of Geography, which placed the economic world and processes of modernisation centre-stage in the discipline. One of the tasks of feminist geography in the immediate future, then, must surely be to break with the dominance of economic explorations in our thinking about space and place. Again, this is a point which we pick up in the conclusion to this volume where we talk about bodily geographies.

READING A: **The Asian miracle, *New Internationalist*, 263: 12–14 (January 1995).**

The walls around the Export Processing Zones, the EPZs, are often topped with barbed wire. The places resemble huge labour camps. Inside them you find the women who were the linchpin of the first phase of the industrialisation process of South Korea, Taiwan, Singapore and now South China and Indonesia.

The women come here when it's boomtime, 'to perform their national duty' and work in the factories. But come 'recession' or 'relocation' and the welcome mat gets yanked in. They are reminded of their responsibility to home and children.

These women are still regarded – often even by themselves – as a secondary or subsidiary work force. In Hong Kong – where nearly all low wage production has moved into EPZs of South China – uneducated women workers in their forties and fifties have been left behind as employment shifts to financial services and communications.

The EPZ economy is based on employing cheap labour for the assembly of high volume standardised components. Such work is seen as particularly suited to women. Since the 1960s young women have been employed in EPZ factories on a large scale. They comprise the majority of child labour, often spending most of their teenage years in sweatshops making plastic toys or garments. By the 1970s they had moved on to more sophisticated assembly lines – particularly electronics and pharmaceuticals. By 1982, of the 62 617 workers employed in EPZs in Taiwan, 85% were women, three quarters of them under 30 years of age. The *Investor's Guide* in Taipei trumpets the advantages of hiring women workers: wage rates fixed at 10% to 20% below those of men (and as much as 50% in some Korean firms): a malleable work-force that won't cause trouble.

Such hints did not fall on deaf ears. Companies sent tour buses to rural villages to persuade inexperienced young girls to leave home. A system of recruitment from junior high schools was approved by the Education Department – the jobs turned out to be a poor substitute for education. Simple assembly tasks left the women with few transferable skills.

The EPZ jobs do offer some young women a way to escape restrictive social roles. But the zones have their own stringent social controls. The dormitories where the women live, as well as the factories themselves, are tightly supervised. There are few opportunities to socialise with young men. Security officers monitor workers during their non-working as well as working hours. In a typical dormitory there are six women to a room. Each room is made up of single beds or bunk beds and a shared common table or two. There is almost no space for personal belongings.

In the south China EPZs fires have killed hundreds of workers who have been locked into their factories and dormitories. Some of these casualties, most of them women, occurred in illegal 'three-in-one' factories. Here workers live on the top floors with the storage facilities while production carries on downstairs – giving new meaning to the phrase 'living for your work'.

READING B: **EPZs in China, *Connexions*, 44: 19 (1994).**

Since early this year, a number of strikes have been reported in Zhuhai, one of the flourishing Special Economic Zones in the south of China. Most of the strikes took place in factories with foreign investments and occurred as a result of the worker's dissatisfaction with low wages. In early April, 800 workers of the Japanese Canon company, which manufactures cameras, went on strike for three days. The strikers demanded a 50% wage increase. According to the workers their monthly wage was around 500 yuan (US$62.50). In another Japanese company, San Mei Machinery, which produces home electrical appliances, workers have gone on strike at least four times this year. During the most recent strike 2000 workers stopped work. The strike soon spread to two nearby Taiwanese companies which produce shoes. The companies employ a total of 3000 workers. They receive an average wage of 300 yuan which is below the legal minimum of 335 yuan. Workers also complained about the poor living conditions of the factory dormitory. While the workers are provided with meals and lodgings, they have to pay 5 yuan a day for food and 40 yuan per month for lodgings. Workers returned to work after the management promised to provide an extra allowance of 60 yuan to supervisors and 20 to the other workers. However, workers who came from other provinces complained that their temporary residence permits and work permits were taken by the management. If they did not have these permits they would not be able to work in any other factories in Zhuhai. There are now about 2000 factories with foreign investment in Zhuhai. The total number of workers employed by these factories is over 200 000, many of whom are young women coming from poorer provinces such as Guangsi, Huanan and Hubei.

READING C: **Domestics in South Africa fight servility and the sack,**
New Internationalist, 254: 29 (1994).

Margaret Nhalpo has a tough job as a South African Domestic Workers Union (SADWU) organiser. 'I think people are taking revenge on domestic workers' she says, referring to increased violence against them. She recalls the woman who visited the Union office, terrified that her employers wanted to kill her, though she had no firm evidence. Some weeks later she was found shot dead in the street. Margaret has scores of reports of rapes during 1993, but a combination of victims' terror and lack of police interest means the culprits always get away. She worries about the vulnerability of domestics to intimidation as the election campaign hots up.

The dawn of the 'new South Africa' is bringing changes, albeit slowly. The basic Conditions of Employment Act affords domestics limited statutory protection for the first time and is seen by the union as a 'foot in the door'. The SADWU office in Cape Town reports calls from more enlightened or image-conscious employers, checking out the Act's implications. Other bosses have greeted the legislation with a wave of dismissals. In Natal SADWU says hundreds of their members have been sacked – and the Union's membership represents only a tiny fraction of this atomised workforce.

'People should pay for the hours they can afford' says Laura Best, from Black Sash, the human rights organisation. She says the character of domestic work is changing. Fewer whites are employing servants to enhance their status and fewer domestics are employed on a live-in basis. 'Most newly built houses in the suburbs don't even have servants' quarters', she says.

This view is shared by Florie de Villiers, SADWU's national president, who hopes lost jobs will be replaced through the new government's reconstruction programme. She also acknowledges that the union has had a particular problem dealing with domestics who are often especially badly paid. 'Domestic workers must have the same protection as all other workers', she insists.

READING D: **Johnston, L. and Valentine, G. 1995. Wherever I lay my girlfriend, that's my home. In Bell, D. and Valentine, G. (eds),** *Mapping Desire.* **London: Routledge, 99–113.**

... The word 'home' has multiple meanings ... shelter, hearth ... heart, abode and paradise... But although the home may be a more or less private place 'for the family', it doesn't necessarily guarantee freedom for individuals from the watchful gaze of other household members ... .

Lesbians living in (or returning to) the 'family' house, who haven't 'come out' to their parents can find that a lack of privacy from the parental gaze constrains their freedom to perform a 'lesbian' identity 'at home' ... it is a location where their sexuality must often take a back seat ... .

The home can therefore be a site of tension for women who identify as lesbians – a place where the ideal of the home as a place of security, freedom and control meets the reality of the home as site where heterosexual family relations ... restrict the performance of a lesbian identity ... .

Tensions between parents and a daughter's lesbian identity can resurface even when she has fled the heterosexual nest. Having a home of one's own may allow a woman enough control over the space to express her sexuality in the physical environment but it doesn't necessarily guarantee freedom from the prying eyes of parents, relatives and neighbours ... .

One way to take the tension out of [visits] is to change the performance of the home according to the identity of the visitor. Whilst some women 'de-dyke' the house completely, others make more subtle changes depending on the level of discomfort likely to be expressed by visitors or experienced by the occupants....

The privacy of the home is not always the same thing as privacy from the neighbours. Prying eyes over the garden fence, eavesdropping through badly sound-proofed walls, and the efficiency of local gossip networks can expose...couples to neighbourhood surveillance. Usually this evokes nothing more than a few snide or petty remarks but occasionally lesbian and gay homes can become the target of hate campaigns ... .

... a lesbian home may become almost a prison. Or a very static and stifling place to be...

READING E: Wilson, F. 1993. Workshops as domestic domains: reflections on small-scale industry in Mexico. *World Development,* 21(6), 67–80.

Workshops as extensions of the domestic domain

The knitwear shops of Santiago, like many other enterprises of the region, began by adopting a household model of labour relations. The labour process in the knitwear workshops was so organised as to bear some resemblance to that of the subsistence producing household. Loom work, performed by one or two males, is strictly segregated from the sewing room of the women. Men start the labour process by knitting the cloth (parallel to men's production of the basic subsistence crop, maize); the older, experienced women cut and machine sew the cloth into garments (parallel to the wives' processing of grain to make tortillas); and recently recruited young women tack, press and finish the garments (parallel to the menial household tasks carried out by daughters). Levels of reimbursement reflect position in the labour process, with the young women entrants paid virtually nothing but aspiring to become higher-paid machine workers next in line to the male loom knitters.

The presence of male employees does not contradict the claim that workshop production is associated with the domestic domain. Not only has sex segregation been strongly enforced, with the owning wife on hand to keep men out of the sewing room, the terms of men's employment have been significantly different from women's. The majority of men working the looms intend to become independent producers.

For young women, employment in a workshop can be considered as providing protection within a surrogate family until marriage ... .

Workshops opening up in a small town such as Santiago in the 1960s presented opportunities for young women to see themselves in a different light from the contradictory images given their mothers' generation, first as 'child brides' and later as 'suffering mothers'. Women fought for better conditions and greater respect in the workshops. At the same time they tried to negotiate with parents and brothers for a little more freedom and autonomy, at least for the right to spend part of their wages on themselves. They demanded a 'youth' between childhood and marriage and they actively engaged in discussions with prospective husbands as to a just financial settlement after marriage. After leaving workshop employment the former workers tried to pass on to their younger sisters, or later to their daughters, what they had learned.

When women found ways of entering wider overarching social relations through workshop employment, they grasped the chance of depicting their own position as women and as workers in a new light. Wage employment under the guise of household relations raised similar questions at home as at the workplace. No matter how much the workshop might consciously pattern themselves on domestic or household models offering 'proper' control and protection of young women, they could not suppress the development of their consciousness as women and as workers.

The consequences of the rise of workshop production in small centres such as Santiago at the present time are ambiguous: many contrasting assessments can be drawn. On the one hand, it can be argued that when capital directly invades and subverts the domestic domain and women's status as family members is taken away, then women are in danger of becoming 'unprotected' by the law. On the other hand, through the expansion of waged work, women are given the opportunities to enter worker solidarities and to experience far greater sense of collectivity with their companions.

CHAPTER SIX

Feminist geographies of environment, nature and landscape

GILLIAN ROSE, VIVIAN KINNAIRD,
MANDY MORRIS AND CATHERINE NASH

6.1 Introduction

This chapter examines three related terms: nature, environment and landscape. Each of these terms has its own particular meanings, but these meanings also intersect; 'nature' is often used synonymously with 'environment', for example, and 'landscape' can be a particular way of understanding 'nature'. The first aim of this chapter is therefore to clarify the meanings of, and the relationships between, these three terms.

Feminists have explored aspects of 'nature', the environment and landscape at some length, whilst feminist geographers have tended to focus on particular aspects of these issues. The environment has often been studied in terms of the built environment, for example, and Section 6.2 of this chapter summarises some of this work. The environment is also often understood as referring to a 'natural' world, however, and attitudes towards, and interactions with, the apparently natural environment have also been examined by feminist geographers; Section 6.3 looks at this work. Both these kinds of studies of the environment show how gender – and class and race, among other social identities – is reproduced through dominant understandings of what the environment is or should be. Both also suggest that the environment, even the supposedly 'natural' environment, cannot be understood without considering the cultural, political and economic processes through which the environment becomes caught up in power relations. One of the ways in which feminist geographers have explored these kinds of issues, especially the power-ridden cultural processes of interpreting the 'natural', has been by looking at 'landscape'. Landscape is one way in which the natural has been made sense of by Western cultures, and Section 6.4 examines feminist analyses of this particular way of seeing.

These analyses by feminist geographers of the environment, 'nature' and landscape argue that the three terms mobilise certain assumptions about gendered identities, and that their deployment when environments are being designed or managed, or landscapes are being seen, can then reproduce those assumptions. The second aim of this chapter is to explore these arguments in some detail. The chapter aims to show just how feminist geographers have argued that the design of the built environment, ideas about the natural environment, and images of landscapes, all in particular ways reproduce specific definitions of gendered difference. Within this broad area of agreement, however, there are different interpretations of exactly what kinds of differences are reproduced by environments and landscapes. Feminist geographers working in this area, regardless of their particular object of interest, echo in their analyses the different emphases found in much feminist geography (see Chapter Three). Some studies focus almost exclusively on gender roles and relations, while other studies stress the intersection of gender with other kinds of social differences. All the sections of this chapter explore work which draws on both of these approaches. The chapter also considers other differences among feminists. The second section shows that discussions among feminists about the natural environment also involve power relations; some Western feminists have suggested that Third World women are more sensitive to environmental issues, but this entails a number of assumptions about femininity in the Third World which many other feminists have challenged. The chapter therefore also explores the need for feminists to consider their own interpretive practices, and the work of feminist geographers on the idea of landscape provides some suggestions about how Western feminists can problematise their views about the environment.

The third aim of this chapter is to show that, if gendered differences can be reproduced through certain understandings of the environment, 'nature' and landscape, then much feminist work has been devoted to challenging those understandings and their associated practices. Feminists have argued for the redesign of the built environment, for different approaches to the 'natural' environment, and for different ways of seeing landscapes. Each section of this chapter also examines these strategies of resistance.

6.2   The built environment: an introduction

As geographers, we are taught to understand the environment in terms of the form of our physical surroundings. One aspect of the environment thus understood which feminist geographers, among others, have focused on is the built surroundings through which we live our everyday lives. Feminists have looked at the houses and streets, corner shops and shopping malls, piazzas and monuments, gardens and offices, car parks and pubs, parks and bingo halls, in which men and women socialise, shop, eat, sleep and work. These elements of the built environment have been studied in terms of the design of individual components such as a house, but also in terms of the integration of several components, for example the streets, houses and gardens of a housing estate. And this work has looked at these issues in both historical and contemporary contexts.

The approach of many feminist geographers to the design of the various elements of the built environment can be neatly summarised by quoting the subtitle of a book written by a feminist design collective called Matrix:

Women and the Man-Made Environment (Matrix, 1984; see also Roberts, 1991). Feminist geographers – as well as feminists working in other disciplines and professions concerned with the built environment – have pointed out that the built environment is almost always surveyed, planned, designed and built by men, and, more importantly, that patriarchal assumptions about gendered identities are articulated through all these processes (Greed, 1994). The built environment quite literally constructs gender. Feminist geographers have thus paid attention to the assumptions about gender which certain kinds of built environment design make. The first sub-section explores this work, and also shows how other feminists have demonstrated that other kinds of social identities can also be reproduced through the meanings of the built environment. The second sub-section looks at some examples of feminist design practice which try not to reproduce dominant gendered identities. This sub-section raises some important questions about differences among women, however, by suggesting that the same design strategies may not be equally relevant to all women.

Feminism and the design of the built environment

Many feminists have argued that the built environment is man-made, both literally in who designs and builds it but also in terms of the assumptions about the social identity and behaviour of the people who will live and work in the environments being designed. These assumptions are often based on dominant views of what appropriate behaviour is, and these assumptions are often highly gendered. Views about what men and women should be doing get translated into views about how the places they inhabit should be shaped. Feminist geographers interpreting the built environment in this way are thus using the second of the ways of understanding gender discussed in Chapter Three: gender as a role.

As an example of the gender assumptions made by dominant forms of housing design, consider the space of the kitchen. One of the most time-consuming elements of the unpaid domestic labour that women are expected to perform in households is cooking. Cooking is still overwhelmingly a feminised task and, as Roberts (1991) argues, it is a task replete with connotations of service. It is a task which the designers of houses and apartments, therefore, almost always give its own space, and the interior layout of many kinds of houses can be understood in relation to the gendering of this task. The typical suburban house in a Western city can be interpreted in this manner. As Matrix (1984) point out, most such housing is built for nuclear families. It assumes that the people living in such houses are a mother, a father and one or more children. The kitchen in such houses is often a small space, as if what goes on there is not very important, and there is rarely enough room for two people to work in it. Just who that one person in the household is assumed to be becomes obvious when, for example, we look at advertisements for fitted kitchens, which rarely show men cooking (although they may show men fitting the kitchen cabinets); or at recipes and cooking tips which overwhelming appear in women's magazines and not men's (although professional celebrity cooks are often men). The kitchen in this domestic setting is a feminised space. Another example can be drawn from the Caribbean. Traditional Caribbean housing design did not include a separate space for cooking inside the house; instead, cooking took place in a rudimentary shelter built outside the house, with stone walls and a galvanised sheeting roof. The cooking area was always hidden from public view, and was usually very cramped and basic – cooking was done over a fire, for example. In modern housing, however, the layout follows Western tradition. The kitchen is inside the house, and is small and at the back of the building. Kitchen space is thus very often designed as a separate space for women's unpaid work because it is assumed that only women will work there.

The design of many post-war housing estates often also makes certain assumptions about gendered difference, again by constructing a space which attempts a gendered segregation of activity. Many women living on such estates either want or need to find paid employment, but the design of estates does not support this. For example, housing estates are rarely provided with creche facilities, on the assumption that childcare takes place in the home, by the mother. Housing estates are often distant from workplaces in a city, so that even if affordable childcare can be found, women who have children to look after find it difficult to travel to a paid job. Because of this lack of adequate childcare, many women with young children want part-time paid employment, but public transport often caters most efficiently for those who work full time. As well as the design of individual living units, then, the design of housing estates also articulates particular assumptions about who will be living in this environment and what they will be doing there.

Moreover, there is extensive research to suggest that the design of housing estates and shopping centres rarely takes the particular concerns of women into consideration. For example, many women are less mobile than men: they may not have access to a car, they may have to take children with them when they leave their home, and they may be carrying heavy shopping. Yet shopping centres rarely offer creches to leave children in safely while their mother shops; city centres are often full of subways, escalators and lifts which are difficult to negotiate with shopping bags or a pushchair; public toilets are often cramped spaces with long queues and no nappy-changing facilities. This is despite the fact that spaces such as department stores and shopping malls are designed as public spaces where women are seen as the primary shoppers and consumers.

ACTIVITY

Visit a nearby shopping mall or centre and try to decide whether it is a 'masculinised' or 'femininised' built environment. What are the criteria on which you base your decision? Is this a segregated space? Is it a safe space, and for who? Are there assumptions about who will be caring for children? Is it a space where children are catered for? Are there different assumptions about gender being made in different spaces of the mall or centre? Are other aspects of social identity assumed by this environment? How would someone in a wheelchair deal with its design, for example? Is the attempt to describe such places as simply gendered not complex enough?

Moreover, the design of the urban built environment neglects the worries many women have about the danger associated with particular locations. Valentine's (1989) discussion of 'The geography of women's fear' (see Chapter Three, Reading C) found that many women avoided certain kinds of urban locations all the time: for example, certain alleyways or subways, certain parks or certain pathways among housing. Other locations they avoided at certain times: high streets at pub closing times, multi-storey car parks at night, empty trains at night. They were scared of these locations because they were all spaces where they felt isolated and vulnerable to attack. Valentine (1992) has also explored the way in which certain spaces are known as dangerous and the way women build up their own understanding of the built environment in terms of safety and risk.

ACTIVITY

Read the following poem by Bronwen Wallace. What sorts of criteria does she use to map a geography of women's fear? Are the effects of this fear confined to the public spaces of streets and pavements?

TO GET TO YOU

It's never easy.
Even the effort of a few steps
from the bedroom to the kitchen, say,
or a few muscles, opening my eyes
to find you, still there in bed beside me
is an act of magic or faith,
I'm never sure which.

All I know is that it's learned
by doing, over and over again,
like any other trick,
until you don't need to think about it.
Like now. Like the way I'm walking home
to you through this city I've learned to accept
as the only kind there is: five o' clock,
night coming down and rain
just hard enough
to make the crowds on the corners shove a little
when a bus finally splashes to the stop.
Outside a restaurant, two men shake hands
and a little boy holds his father's
as they watch a toy airplane turning in a shop window.
It could be anywhere. But what I want you to notice
are the women. They are wearing white nurses shoes,
or dirty sneakers or high-heeled boots.
They carry briefcases and flowers, bags of groceries
as they hurry home to husbands and kids,
lovers, ailing parents, friends.

We all have the same look somehow.
See: over there by the bank
how that stout woman lowers her eyes
when she passes that group of boys,
how her movement's mimed
by the blonde, turning her head
when a car slows down beside her.
Even the high-pitched giggle of the girls
in that bunch of teenagers is a signal
I've learned to recognize. Tuned in
by my own tightened muscles, jawline or shoulders.
In fact, you might study the shoulders.
The line of the backbone, too; arms and hips,
the body carried
like something the woman's not sure what to do with.

I've already told you that this is an ordinary city.
There are maps of it and lights to show us
when to walk, where to turn.
What I want you to know is that it isn't enough.

On a trip to Vancouver once
I discovered clearer landmarks. Red ones,
sprayed on sidewalks all over the city.
They marked the places
where a woman had been raped,
so that when I stepped out of a coffee shop
to find one on the pavement by the laundromat
geography shifted.
Brought me to the city I'd always imagined
happening in dark alleys, deserted parking lots,
to somebody else. Brought me home in a way,
no longer the victim of rumours or old news,
that red mark planted in the pavement
like the flag of an ancient, immediate war.

I used to hope it was enough
that you were gentle
that I love you,
but what can enough mean, anymore
what can it measure?

How many rapes were enough
for those women in Vancouver
before they got stencils and spray paint
made a word for their rage?
How many more until even that word
lost its meaning
and the enemy was anything that moved out there.
Anything male, that is.

How can any woman say
she loves a man enough

when every city on the planet
is a minefield
she must pick her way through
just to reach him?

It's not that we manage it, though.
It's that we make it look so easy.
These women wearing their fear
like a habit of speech or movement
as if this were the way
the female body's meant to be.
The way I turn the last corner now,
open the door to find you
drinking wine and reading the newspaper,
another glass already filled
and waiting on the coffee table.

When I turn on the hall light
the city will retreat into the rain,
the tiny squares of yellow
marking the other rooms
where men and women greet each other.
It's a matter of a few steps,
magic or faith, though it's not that simple.
The way the rain keeps watering the cities of the world.
How it throws itself against our window,
harder, more insistent,
so that we both hear.

Bronwen Wallace
(from *Full Moon: An Anthology of Candian Women
Poets* ed. Janice LaDuke and Steve Luxton,
1983, 175–177)

Many feminists, then, have argued that the built environment is designed on
the basis of many assumptions about gender. Its layout assumes that certain
people will undertake certain tasks and not others. These assumptions are reinforced by dominant ideas about how different city spaces should be used, and by
the knowledges women bring to bear on their use of city spaces. While much
work has focused on the experiences of women in these value-laden environments, these feminists are not mapping a geography of women only. Many
women find contemporary urban spaces difficult to live in so that contemporary
models of design have been described as constructing masculinised public spaces
where only men feel safe and can confidently use the space. Geographer Jan
Monk (1992), for example, has argued that public space in the Western city is
built by and for dominant masculinities: men who do not need to worry about
food shopping, kids or being attacked. As a small but telling example of this, she
points to the number of statues of 'great' men which can be found in the public
spaces of cities. What is being studied by these feminists, then, is a particular
expression of the patriarchal construction of relations between men and women.

However, many feminist studies of the built environment have
that the built environment is more complex than this. These are fer
closely aligned with the third kind of understanding of gender c
Chapter Three, and they suggest that the built environment is no
by assumptions about gender alone. The grounds for this argument are perhaps already obvious. For example, the gendering of housing for nuclear
families assumes not only certain gender roles, but also heterosexuality as the
norm: it assumes that the family living in it consists of a man, a woman and
children. Moreover, the difficulties a mother with a child in a pushchair might
encounter in a shopping centre might also be experienced by anyone who is
less mobile: some elderly people, for example, or people with certain disabili-
ties. Again, some women may have the resources to use cities easily, always
taking taxis, for example, or always using a car with a mobile phone. Women's
fear of urban spaces, then, is nuanced by other elements of social difference,
including physical ability, age and class, to take just three examples.

The article by Gill Valentine on 'The geography of women's fear' (Valentine, 1989) provides a systematic account of the complex ways in which some
different social identities intersect to construct dominant ideas about urban
environments. Although she was studying something almost all women feel at
one time or another – fear of attack – she argued that not all women see the
city as threatening attack in the same way. Other social identities inflected
what is perhaps one of the most widely shared experiences of women, so that
there was not just one geography of fear in the large town in southern England she was studying, but several. Her research showed, for example, that
white women in that town were more frightened of entering housing estates
with a high percentage of black residents than they were of entering housing
estates with mostly white residents. Clearly, racist fears about black violence
were shaping white women's geography of fear. Similarly, Valentine also discovered that class position made a difference to fear; middle-class women, for
example, were more scared to be in working-class areas than in middle-class
areas because working-class areas were perceived to be more rowdy, and
more risky to be in, than the environments middle-class women were used to.

It is also important to point out that, just as not all women feel equally
afraid of certain kinds of places, nor do all men feel equally safe in all kinds
of places, and this too is a consequence of different social relations intersect-
ing to create different kinds of gender identities: in this case, different kinds of
masculinities. Gay men, for example, in parallel with lesbian women, may feel
frightened of 'gay-bashing' and that fear may restrict their use of urban space.
Black men may fear racist attacks. And in some households it is the men who
undertake the childcare, and they too find the same difficulties in coping with
the urban built environment as their female childcaring counterparts. The
meanings with which the built environment resonates are gendered as both
masculine and feminine, then, but these arguments suggest that the gendering
of urban locations is mediated also by the relations of race, sexuality, class,
age and physical ability as well.

The design of the built environment clearly assumes that some kinds of activities will be undertaken by women and others by men. In this sense, the built environment is gendered. It is also gendered because women and men often decide how to use the environment on the basis of their understanding of their own gendered identity. However, as this section has shown, gender is not the only social relation to be reproduced through the meanings assumed by, and given to, the built environment. So too are class, sexuality and race, to name but three other aspects of social identity.

Feminist challenges to the so-called man-made built environment

For many feminists, the built environment is an important site of the repro-duction of dominant ideas about masculinity and femininity. It is not surprising, then, that many feminists have also attempted to challenge con-ventional ways of designing built environments so that women's particular needs are catered for. Such challenges have a long history. Dolores Hayden (1981), among others, has pointed to the many plans drawn up by women (and by men) in the nineteenth century for housing which did not place the nuclear family in its own house and reinforce in its bricks and mortar the gen-dered division of domestic labour. Many women designed housing for young single women who wanted the independence of working and living in large cities alone but safely; many designed housing estates with shared kitchen and laundry facilities to liberate women with families from domestic tasks. These latter experiments are a reminder of the importance of not considering gender in isolation from other social relations, however, since many of these designs were intended to liberate middle-class women from household management and they relied on the labour of working-class women as cooks and laun-dresses to achieve this.

In cities now, perhaps the most common tactic to make environments more women-friendly is to campaign for certain spaces to be more sensitively designed and better-lit, so that women feel safer in them. Feminist geogra-phers Gerda Wekerle and Carolyn Whitzman (1995) have examined many such design projects in their book *Safe Cities*. 'Safe city' projects involve locally based alliances of interested parties which aim to combat the fear of crime in urban neighbourhoods. Although often co-ordinated by local govern-ment, these projects assume that it is local people who are the experts in being able to identify what kinds of places induce fear, and in suggesting what could be done to improve them. Wekerle and Whitzman suggest that women often make strong contributions to safe city projects precisely because they often experience fear when using their local neighbourhood; 'women who live in cities are engaged in an ongoing situational analysis of the environments of daily life' and are thus experts at the issues with which safe city projects are concerned (Wekerle and Whitzman, 1995: 4).

Wekerle and Whitzman provide many examples of women organising to change the design of their neighbourhoods by addressing issues such as street layout, lighting and signage, the provision of help points, and spaces of entrapment. Women have also contributed to the design of individual buildings. An example of this is the design of the Jagonari Centre in Whitechapel (Open University, 1991). This is an educational and welfare centre for local women, and it is situated in the heart of East London's large Bengali community. Both its design and the process through which its design came into being are innovative. The previous section argued that designs for the built environment very often make assumptions about what kinds of people will be doing what kinds of activities in the environment being designed. The women who wanted the Centre deliberately looked for an architects' practice which would listen to them, since, like all inhabitants of a built environment, they had developed their own ideas about what a good building for them would be. The chosen architects of the Jagonari Centre – a feminist practice – came to the project with as few preconceptions as possible. They wanted to listen to, and to work with, what the eventual users of the Centre wanted. The design process was participative, then; it did not implicitly assume a certain kind of user. As a result, the building is sensitive to the needs of the particular group of women who visit it for advice or for classes or for social activities. For example, the building is very secure. Racist attacks are common in this area – all of the women on the Centre's Management Committee have been attacked by racists – so the windows have grills on them and the front door is solid. The grills are fashioned to make reference to the architectural traditions of the many Asian cultures of the women who use the Centre, and the colourings too – pastel turquoises and greens – are drawn from these non-Western traditions. However, the building is faced with brick so that, like the women themselves, it is also part of the tradition of Western urban environments too. The canteen is large and functional, built so that the large pots which Bengali cooks use are easy to move and to wash: the sink is built on the floor. The creche is integral to the whole building, and is a space where children can play safely but without being too far from their mothers. The Jagonari Centre, then, is an example of a built environment which tries to be sensitive to the particular group of people using it. Its design was a collaborative process which depended on the women's own knowledge about their urban environment and their own specific identity. Its spatial organisation and its aesthetic style both reflect a particular gendered, classed and racialised identity which is very different from that usually assumed by the built urban environments in the West. It is a good example of work by a feminist practice which aims to make the built environment more sensitive to the diversity of the populations which inhabit its spaces.

Other feminist work has looked at how women themselves give their own meanings to the built environments they encounter. Just as Valentine (1992) argued that women develop their own knowledge about what sorts of urban environments are dangerous, so feminist geographers have suggested that women can also give positive meanings to certain places which allow women to live more in the way that they want to. This is obviously the case in campaigns to make cities safer for women. For example, when a serial killer was murdering women in Leeds in the early 1980s, women were told by the police to stay indoors; but some women's groups retorted that it wasn't women who should leave the streets but men, since it was a man who was making the city unsafe, not a woman. Women can remake the meanings of existing built environments in less overt kinds of ways too. Some feminist geographers have argued that this remaking may happen, to a degree at least, for some of the women occupying gentrified housing. Gentrified housing is usually older housing located in inner city areas of Western cities which has been 'done up', often by middle-class professionals. Both Bondi (1991, 1992b) and Rose (1984) suggest that gentrifiers are often women who move to the inner city to evade some of the gender assumptions prevalent in suburban housing. Gentrifiers often have non-traditional household structures; they may be single mothers, for example, looking for a somewhat alternative lifestyle close to the city centre, or lesbian women (and work in North America has also suggested that middle-class gay men often choose to live in gentrified areas, for similar reasons). Bondi (1992b: 167) thus comments that 'gentrification offers scope for reworking images of femininity and masculinity in ways that encompass diverse and non-traditional forms'. Although Bondi also insists that there are limits to this process, going on to say that gentrification 'does little to disturb existing gender relations', her work does suggest the possibility of women remaking the gendered meanings of the already-existing built environment.

SUMMARY

- Gendered identities (which are also sexualised, racialised, classed and so on) can be reproduced through the assumptions about social identity made by the design of the built environment.
- The built environment is given meaning both through dominant ideas of who should be where but also by the users of the environment.
- Particular women have attempted to redesign and re-imagine specific built environments based on their own perceptions of what they want.

6.3   The environment: an introduction

Feminist geographers are beginning to pay some attention to another understanding of the environment: the 'natural' environment. Feminists have critically engaged with dominant definitions of the term *nature* (see Box 6.1), and feminist discussions of the 'natural environment' depend in different ways on elements of this extensive critique.

Box 6.1   Nature

'Nature' is an extraordinarily complex term in Western cultures. Perhaps two dominant meanings can be detected, though (Livingstone, 1986). Firstly, 'nature' can be used to refer to the essence of something. Secondly, 'nature' can be used to refer to the world in general, particularly the physical world conceptualised as separate from the human world. This chapter is mainly concerned with the latter emphasis; but feminists have clearly engaged with the former when they debate the 'nature' of masculinity and femininity, or when they question the essentialism implicit in notions of 'nature' when applied to social identity (see Chapter Three); the debates around environmental feminisms have certainly drawn on both meanings. 'Nature' as a reference to the physical world has historically been feminised in Western discourses, and feminist writers stress how this feminised 'nature' is often understood as separate from, and vulnerable to, a masculinised understanding of culture. More recent discussions have suggested that 'nature' can be seen as the Other of 'culture', and, as an Other, 'nature' has an ambivalent relation to 'culture'. Thus feminised 'nature' can be seen as nurturing and plentiful, bounteously reproducing the fruits of the earth; or she can be seen as a mysterious and uncontrollable force, threatening the achievements of culture and civilisation. This of course parallels the two ways in which women are most often represented in cultural discourses, as either madonna or whore, and influential essays by Rosaldo (1974) and Ortner (1974) argued that women's oppression could be explained by their association with the 'natural' and by men's association with what was defined as the more valuable cultural.

The term 'natural' often implies something separate from the human, with its own processes distinct from those of society, economy, polity and culture. Feminist geographer Joni Seager (1993) has explored the understanding of 'nature' as separate from the 'human' in the context of environmental issues. She suggests that when the media report on environmental issues, the environment is very often conceptualised purely in its physical form: pollution is described as 'having an impact on' the environment, for example. She also suggests that many of the earth sciences follow this separation of a 'natural environment' from human life. Her work explores some of the implications of such a separation. First, a focus on the environment as a natural, physical entity often leaves questions regarding causality on the sidelines. Consequently, questions about human agency, that is, the fact that social and economic forces are, at least partially, responsible for the state of the environment we see around us, are 'placed a distant second, if they are realised at all' (Seager, 1993: 2). Secondly, our universal belief in the ability of the 'natural sciences' to understand that 'natural environment', again, does not allow us to question the human-related reasons for severe decline in environmental quality. And if the natural environment is thus constructed by both the media and by the sciences as quite separate from human processes, there seems little possibility for developing a feminist critique of current human–environment relations. The framing of environmental problems in purely physical terms is 'barren territory for feminist analysis: there is no feminist analysis of the chemical process of ozone disintegration; there is no feminist analysis of soil erosion, of ground water pollution, of the acidification process which is killing forests and lakes throughout the industrialised world' (Seager, 1993: 3).

Nonetheless, Seager insists that feminists should step into the analytical vacuum left by both the scientific community and the ways in which environmental issues have been popularised. Seager argues that feminists ask questions about human agency and therefore are able to offer an understanding of the 'natural' environment which relates that environment to social, political, economic and cultural processes. Feminism can look at the ways in which social power relations – for example, the struggle of vested interest groups – construct particular understandings of the environment and have particular material impacts on the environment. In her own work as a human geographer, working in an American university, Joni Seager has been influential in setting an agenda for a feminist geography of the environment. She says that the aim of her research is to deconstruct the workings of institutionalised power by asking 'who' is responsible for environmental crises. As a feminist, an important aspect of the answer to that question is a careful assessment of men's and women's relationships to the institutions which mediate our relations with the environment.

Seager's work suggests that feminist geographers can contribute a great deal to analyses of the environment. In fact, much of the work discussed in this section of the chapter does not originate from geographers. However, we recognise that since environment is one of the central themes of geographical enquiry, we need to begin to construct feminist geographies of the environment which offer a critique of the dominant scientific interpretation of 'environment as physical processes' which structures geography as an academic discipline. We can provide a critical view of human–environment relations that questions the political, social and economic structures of environmental processes. This aim is not so very different from the way in which feminists have contextualised the built environment in relation to the social, economic, political and cultural values in which it is embedded. There is another parallel between the work of feminist geographers on the built environment and those feminists working on environmental issues, too: their understanding of gender. Like feminists examining the built environment, those examining the 'natural' environment have looked at both gender roles and gender relations as their primary analytical focus, and at the intersection of gender with other social differences. This section explores both these approaches in the work of feminist environmentalists.

This section also raises another issue, however. In looking at different feminist understandings of gender, this section begins to examine some of the ways in which feminists themselves can be divided by social difference. In particular, this section looks at some of the differences which have emerged between some (though not all) Western feminists and some (though not all) Third World feminists.

'Nature', science and gender

In order to understand environmental feminisms, we must begin by discussing two fundamental aspects of feminist inquiry: the nature/culture dualism and the feminist critique of the dominant scientific presumptions that shape our social, political and economic lives.

As Box 6.1 indicated, 'nature' has historically been gendered as feminine in Western discourses. Feminists concerned about the environment have seen this aspect of the nature/culture dualism as central to understanding both historical and contemporary attitudes towards the 'natural' environment (Plumwood, 1992). Carolyn Merchant (1981) and Ludmilla Jordanova (1989), among others, have examined how dominant forms of science since the seventeenth century have assumed that the scientific gaze at the world looks at a feminine 'nature'. Merchant (1981) argues that because greater value was placed on this scientific knowledge than on 'nature', a particularly exploitive approach to the environment was produced. Indeed, some feminists have argued that 'nature' is universally devalued (Ortner, 1974) and have gone on to examine the consequences of that for women. Vandana Shiva and Maria Mies (1993), for example, draw parallels between the ways in which modern science views women's experience of childbirth and the way in which it controls 'nature'. For the former, women's bodies, they argue, are merely a vehicle through which the scientific 'experts' deliver a new human life. With respect to 'nature', Shiva and Mies use the example of the hybridisation of plants in order to speed up the process of reproduction of food-bearing plant life because 'Nature's ways of renewing plants are dismissed as too slow and "primitive"' (Shiva and Mies, 1993: 28). Thus the equation of femininity with 'nature' has been used to oppress women by suggesting that women themselves are more 'natural' than men. Moreover, the effects of the way this dualism structures dominant notions of scientific knowledge have also been to 'naturalize domination' (Reuther, 1975); identities of both the dominant and subordinated groups are constructed as 'natural' by this science, and therefore also as unchallengeable (Plumwood, 1992).

This feminisation of 'nature' (and 'naturalisation' of femininity) went hand in hand with a masculinisation of 'science'. The links between Western scientific conceptualisations of the environment and Enlightenment-based meanings of scientific and technological progress are common themes throughout the environmental feminist literature. Most base their discussions on feminist scholarship which critiques mainstream science's claim to be a universal, value-free pursuit of truth (Braidotti et al., 1993). Feminist critics point to the emergence of a Western science which has developed since the Enlightenment period and has become firmly established within our political, social and economic institutions, and describe it as a particular kind of masculinity:

> The very definition of 'the scientific mind' is coterminous with rationality, masculinity and power. The scientist as model for subject of knowledge is therefore defined in a set of hierarchical relations to others: the non-scientist. Feminists have criticised scientific discourse as an account of the world that systematically devalues every category that is 'other' than the male, Western, bourgeois self: women, children, other races, foreign cultures, lower classes, handicapped people and nature. (Braidotti et al., 1993: 31)

As Chapter Three argued, this Western, male-dominated view of science is the dominant ideology through which geographical academic research is conducted and through which students of geography are traditionally taught. It is possible to suggest that physical geography in particular remains committed to a belief in objective and universal knowledge, but feminist geographers have only just begun to consider the possibility of a feminist physical geography. The possibilities are intriguing, and you may like to consider them at the end of this chapter when you have read about the range of feminist critiques that have been directed at different conceptions of the 'natural environment'. We can also note here that feminist critiques of the production of 'scientific knowledge' are also part of the postcolonial critiques of Western knowledges discussed at the end of Chapter Three, and we can now see that this analysis of Western scientific knowledge is relevant to a wide range of feminist work.

Feminists concerned about the environment have theorised the masculinity of dominant forms of Western scientific knowledge in a number of ways. Some argue that men, and science, are essentially oppressive and exploitive; others insist that the elision of certain kinds of masculinity with certain kinds of scientific knowledge is a social construction and it can therefore change and be contested. Those arguments are not a focus of concern here, however. Rather, this section now turns to the ways in which feminists concerned about the environment have reacted to the claim that the Western masculinity in some way underpins Western sciences, and the next two sections examine some of those reactions.

Ecofeminism

There are many ways of linking feminism to the environment. This diversity is indicated by the number of terms used to refer to such links: *ecofeminism*, *ecological feminism* or *feminist ecology*. A wide variety of issues are discussed, including Third World development, green consumerism, feminist environmental spirituality and environmental philosophies. According to Cathy Nesmith and Sarah Radcliffe (1993), these debates comprise an innovative mixture of academic and non-academic ways of thinking, including poetry and fiction. Many elements of these diverse positions and debates are informed by different kinds of theory, including, but not exclusively, feminist theories. This section focuses on the feminist theories about gender relations and the environment and examines one of the most important feminist arguments, ecofeminism.

The term ecofeminism was first used by Françoise d'Eaubonne in 1984 to describe women's collective ability to bring about an ecological revolution. She blamed the most immediate threats to our global survival on what she called 'the Male System', which needed to be destroyed by women after which 'the planet in the feminine gender would become green again for all' (d'Eaubonne, 1984: 236). Today, ecofeminism is a major sub-group of the feminist movement; it is largely activist-based and its theory is as complex and diverse as feminism itself (Plumwood, 1992). It contains a range of positions which connect the domination of women with that of 'nature' and stresses the inter-connectedness of feminist and ecological concerns (Braidotti *et al.*, 1993). What holds these diverse positions together, though, is a vision of society beyond militarism, hierarchy and the destruction of 'nature' (Plumwood, 1992). Vandana Shiva and Maria Mies sum up the ecofeminist cause as:

> ... a women identified movement ...[with] special work to do in these imperilled times. We see the devastation of the earth and her beings by the corporate war-riors, and the threat of nuclear annihilation by the military warriors, as feminist concerns. It is the same masculinist mentality which would deny us our right to our own bodies and our sexuality, and which depends on multiple systems of dominance and state power to have its way. (Shiva and Mies, 1993: 14)

Despite this diversity, it is possible to identify two main theoretical positions to be found with ecofeminism. Following Plumwood (1992), we have described these two broad positions as *cultural ecofeminism* and *social ecofeminism*. These are outlined in Box 6.2, although it should be emphasised that this distinction is by no means watertight and the same campaigns and the same feminists can contain elements of both these positions and more.

Box 6.2   Ecofeminisms

Cultural ecofeminism: Cultural ecofeminism stresses the links between women and 'nature' and their joint oppression as a consequence of male domination. Women are seen as having a superior relation to 'nature' which is often taken to be biologically determined. Cultural ecofeminists emphasise a spiritual relation-ship with 'nature' that stresses personal transformation and the (re)empowerment of women and women's values. To solve environmental problems, cultural ecofeminists advocate the creation of an alternative 'women's culture'.

Social Ecofeminism: Social ecofeminists emphasise the social and political aspects of ecofeminism. They reject the biological determinism of cultural ecofeminism, opting instead to view 'nature' as a political rather than a natural category. Social ecofeminists would argue that the entire development of dominant culture and its relationship to 'nature' has been affected by male and other forms of dominance as expressed through the dualism of 'nature' and scientific reason.

Source: Plumwood (1992).

An example of cultural ecofeminism is the book by Vandana Shiva and Maria Mies (1993) called *Ecofeminism*, referred to in the previous section. Their arguments in that book are based on what they see as women's essential knowledge of 'nature' as expressed through the 'feminine principle'. Feminist geographers have also suggested that women are somehow necessarily closer to and more sensitive to 'nature'. Jan Monk and Vera Norwood, for example, have edited a collection of essays about women's relationship to the desert landscapes of the southwestern USA which is close to this position. Although they acknowledge differences between Anglo, American Indian, Mexican–

American and Hispanic women in that region, they nevertheless suggest that 'women of all cultural groups have found in the landscape (both natural and constructed) a source of strength and personal identity' (Monk and Norwood, 1987: 9). The essays argue that the environments of the southwest fascinated the women who lived in that region, and the women in turn cared for the environment there. Monk and Norwood connect this nurturance with women's role in nurturing children, and thus they generalise it across different groups of women. Cultural ecofeminism thus argues that women must be responsible for the solution to environmental problems.

This form of cultural ecofeminism sits uncomfortably with many Western feminists and environmentalists who are concerned with the *essentialism* which cultural ecofeminism and the 'feminine principle' assume (see Box 2.2 for a definition of essentialism). However, as Nesmith and Radcliffe (1993) discuss, many other Western feminists do not see the essentialist connotations of cultural ecofeminism as a problem because their understanding of gender is mediated by their understandings of other social identities. In particular, many Western feminists believe that the West is so overdeveloped that getting in touch with 'nature' is especially difficult there. In contrast, they believe that the Third World remains closer to 'nature', and that women there, because they undertake so much work maintaining household resources, remain close to 'nature'. Jane Jacobs (1994) examines this argument as it is made by some white feminists about Aboriginal women in Australia. But, as both Jacobs (1994) and Nesmith and Radcliffe (1993) make clear, this is a patronising attitude to take. It depends on a long history of the West as seeing itself as more advanced, more cultured and more distant from 'nature' than the rest of the world; it is a way of seeing the Third World as a place where understandings of the natural world are not as culturally mediated as they are in the West. In short, it denies the Third World its cultural practices.

Some ecofeminists, especially social ecofeminists, have challenged this view of Third World women, arguing that it depends on a culturally specific stereotype of those women. By and large, social ecofeminism is marginal to the ecofeminist activist cause and is largely contained to discussions within academia (see Merchant, 1981; Warren, 1994). Social ecofeminists reject the idea that women's closeness to 'nature' is a result of biological destiny. Instead, they favour a closer look at the social and political constructions of gendered relationships with the environment.

The differences in emphasis between cultural ecofeminists and social ecofeminists have a distinct geographical context. Some would argue that the concern over women's essential relationship with the environment is irrelevant for women throughout the Third World because the cultures of the South do not perceive the male/female dichotomy in the same way as women from the North (Braidotti *et al.*, 1993; Shiva and Mies, 1993). As a result, it is not problematic for these women to identify themselves with 'nature' and use this identification as a source of empowerment (Lamb, 1994/95). Sue Lamb exemplifies this argument by citing the case of the Chipko movement in India where women protect indigenous forestry practices by physically embracing trees (see the following case study for more details). This is often represented in the West as a romantic tree-hugging project allying women with the very natural resource they seek to protect. But, as Lamb argues, precisely because women are not seen as closer to the environment in this area of India as they would be in the West, the activities of women in the Chipko movement require immense personal courage as they face the disapproval of men in their families, communities and the state (Lamb, 1994/95).

CASE STUDY: **The Chipko movement, India**

A fight for truth has begun
At Sinsyari Khala
A fight for rights has begun
At Malkot Thano
Sister, it is a fight to protect
Our mountains and forests.
They give us life
Embrace the life of the living trees and streams
Clasp them to your hearts
Resist the digging of the mountains
That brings death to our forests and streams
A fight for life has begun
At Sinsyari Khala.
Ghanshyam 'Shalland, Chipko poet

India's forests are a critical resource for the subsistence of the country's rural peoples because they provide fuel, food and fodder and stabilise soil and water resources. As these forests have been increasingly felled for commerce and industry, Indian villagers – mainly women – have sought to protect their livelihood through the Gandhian method of *satyagraha*, non-violent resistance. In the 1970s and 1980s, this resistance to the destruction of forests spread throughout India and became known as the Chipko movement. In 1974, village women of the Reni forests of the Chamoli district in Uttar Pradesh decided to act against a commercial enterprise about to fell some 2500 trees. The women were alone: the menfolk had left home in search of work. When the contractors arrived, the women went into the forest, joined hands and encircled the trees ('Chipko' means to hug). The women told the cutters that to cut the trees, they would first have to cut off their heads. The contractors withdrew and the forest was saved. The movement spread as more and more villagers throughout the Himalayas began to fast for the forests, guard and wrap themselves around trees scheduled to be felled, saving them by interposing their bodies between them and all the contractors' axes. In 1980, as a result, Indira Gandhi issued a 15-year ban on green felling the Uttar Pradesh forests. Since then, the movement has spread to Himachel Pradesh, Karnataka, Rajasthan, Bihar and the Vindayas and generated pressure for a natural resource policy that is more sensitive to people's needs and ecological requirements. The Chipko movement is the result of hundreds of decentralised and locally autonomous initiatives. Its leaders and activists are primarily village women acting to save their means of subsistence and their communities. (Chipko Information Centre, India, IDOC *Internazionale*, March 1989, reprinted in Rodda, A., 1991, *Women and the Environment*, London: Zed Books)

If we reject the essentialism of the cultural ecofeminists, yet still want to recognise the empowering force which the relationship between women and 'nature' creates, we must look for other explanations. Braidotti *et al.* (1993) support Bina Agarwal's (1992) explanation for women's emergence as the main activists in the Indian environmental movement. She maintains that women are at the forefront of the movement not because they are women, but because they are marginalised in the social, political and economic processes of change in India and therefore have created a link with 'nature'. Thus social ecofeminists understand the connections between women and the environment, not as *essential*, but as *constructed*, in particular places at particular times in particular ways for particular reasons. It is also important to remember the sheer diversity of women's involvement in environmental campaigns. Women's environmental activism is certainly not confined to the Third World. Women are active globally, and the ways in which they understand their connections to the environment are also likely to vary globally.

There is no doubt that ecofeminism is one of the most promising movements within critical environmental thought (Archambault, 1993). Its radicalism and innovatory ways of theorising and protesting are all sources of inspiration. This sub-section has suggested, though, that a measure of caution is required when we evaluate the theoretical principles of ecofeminism and relate them to a variety of action-based movements which are fighting different environmental causes in different localities of the world. Ecofeminist arguments which claim that all women, but especially Third World women, are closer to 'nature' and therefore more natural environmental activists, can be criticised for their essentialism. Such arguments gloss over differences between women, but at another level they also rely on a differentiation between Western women and Third World women which depends on Western understandings of the difference between the West and the rest.

Women, environment and sustainable development: coalitions of difference

As we have already noted, a variety of ecofeminisms co-exist. The previous sub-section examined some of the main differences among ecofeminist arguments, and paid some attention to the essentialist tendencies in some ecofeminist work. This section examines some of the implications not only of understanding specific relationships between women and the environment as contructed rather than essential, but also of thinking about the ways in which different women may have different relations to the environment.

Those feminists who emphasise the importance of differences among women are also represented in the ecofeminist movement. They too insist that women's diverse environmental actions cannot be reduced to a universal, essential 'feminine principle'. Their arguments have already been rehearsed in the previous sub-section in relation to global differences among women. But this is important to remember at the local level too. For example, it is highly likely that the very different kinds of women (and men) involved in campaigning against the Newbury bypass in southeast England in the mid-1990s held a range of different views about their motivations and actions, and their views may be different again from those of other women in Newbury whose environmental activism involves recycling household waste, for example.

ACTIVITY

Agenda 21 in the UK
Consideration of 'global' and 'local' issues now figures widely in environmental awareness due in part to the policies and commitments agreed at the Rio Earth Summit in 1992 and published by the United Nations Conference on Environment and Development (UNCED) as *Agenda 21*. *Agenda 21* aims to address the global environmental problems of today while, at the same time, preparing the world for the challenges of the next century (UNCED, 1992). Section III of the *Agenda 21* document is entitled 'Strengthening the role of major groups'. Here, actions are identified to enhance the participation of different interest groups, from business and industry to trade unions, non-government organisations, local authorities, children and youth groups and women. These latter groups are identified in recognition of the fact that many environmental problems addressed by *Agenda 21* have their solutions rooted in local activities. Throughout the UK, local authorities are committed to adopting a local *Agenda 21* for their area, thereby defining a strategy for sustainable development at the local level. In co-operation with local interest groups, local authorities were aiming to achieve a consensus on a local *Agenda 21* strategy by 1996; they were encouraged to ensure that women and youth are particularly represented in decision-making, planning and implementation processes. In practice the central government's response to the implementation of *Agenda 21* offers little insight into how this can be done (Buckingham-Hatfield, 1994).
Some recent research by Susan Buckingham-Hatfield (1994) indicates the difficulties of ensuring that women are represented in the local *Agenda 21* decision-making process. She criticises *Agenda 21* for assuming that all women, regardless of their age, educational attainment, ethnicity and marital and parental status, share common environmental concerns.

- Contact your local authority and find out about the local *Agenda 21* in your area.
- What consultative procedures took place in your community?
- Were special efforts made to contact women? If so, how?
- What differences among women in your local area might make a difference to their participation in the local *Agenda 21* project? Here you might find it useful to return to Chapter Three and think about the main differences between women which feminist geographers have so far addressed. Are there other differences relevant to this particular Activity that you think feminist geographers have neglected?

- You may also like to compare your answers with those of another student looking at a different local *Agenda 21*. Are there also geographical differences
  between the potential for women's involvement in *Agenda 21*?

Differences among women may thus affect the ways in which women construct links between themselves and the environment. One feminist
environmental organisation which pays a great deal of attention to such differences is the Women, Environment and Sustainable Development (WED)
movement. WED is a network of grassroots environmental groups in Third
World countries. It grew out of a recognition that the Western project of mod-
ernising the Third World was not yielding a significant improvement in the
living conditions for the majority of people (Braidotti *et al.*, 1993). In fact, the
processes of change that result from development have led to an increase in
poverty and an increase both in gender inequalities and in the degradation of
the environment in many developing-world societies. From both a theoretical
and practical point of view, analysing the development and 'nature' of the
WED movement sheds light on the ways in which feminism and environments
can be connected in diverse ways, in the context of development.

In the book *Women, the Environment and Sustainable Development*
(Braidotti *et al.*, 1993) the authors aim to look at the differences between
women, feminisms and various environmental movements and the implications of these differences for building feminist environmental theory and
knowledge (Lamb, 1994/95). Although many environmental issues have
global impacts and require action at the global scale, WED refuses to suggest
that this requirement demands that all environmental feminists act similarly.
Instead, it concentrates on building alliances between the different strategies
of different environmental groups, because it recognises that the ways in
which women identify with, and find solutions to, environmental problems
are very diverse. Braidotti *et al.* (1993: 169) negotiate this complexity by
turning to the notion of 'situated knowledge' discussed in Chapter Two:

> The implication of increasingly sophisticated technologies and new forms of
> domination is that all those concerned to facilitate pro-environmental changes
> have a political and historic responsibility to critically analyse their own posi-
> tion in the wider power structure in order to identify points of leverage from
> their respective position. This could be as a member of the board of an indus-
> trial company, a scientific institution, a citizens movement or as a consumer.
> This strategy of 'situating oneself' is the basis for a new type of micro-scale poli-
> tics, which relies on temporary and mobile coalitions with other social actors or
> groups, not on the basis of identity, but of affinity of world views and a shared
> sense of ecological ethics.

This argument allows for locally specific (and locally diverse) constructions of
women–environment relations, while connecting such local specificities to a
global coalition for change.

Through the WED movement a diverse number of relationships to the
environment are articulated. The discourse of sustainability and the linking of
environment and development at the global level, as for example through the

Earth Summit, has created the space for different groups to be heard. The WED movement uses this space to construct a broad, shared understanding of the environment which is basically a critique of dominant views of development for Third World countries. The WED movement aims to illustrate the urgent need to find an alternative development model for the developing world by focusing on women and their relationships with 'nature' and the environment. But the multiplicity of environmental feminist theory and activism that encompasses the WED movement displays a wide range of views and specific definitions of environment itself. The WED movement is global and yet its component parts are local and construct definitions of environment which are particular to their specific contexts.

Ecofeminism remains a very diverse field of action and debate, then. Its arguments raise some crucial issues about different kinds of feminisms. It also raises some challenging issues for feminist geographers concerned with rethinking the environment in radical ways.

SUMMARY

- Gender identity can be drawn on in different ways to contest dominant ideas about the environment.
- Feminist politics can itself be caught up in the reproduction of powerful ideas about certain kind of femininities.
- Relationships between particular understandings of the environment and specific and diverse femininities can be constructed in varied ways.

6.4   Landscape: an introduction

The previous two sections have shown just how diverse studies of 'the environment' can be. This section explores another aspect of its study: landscape. Landscape is a key way of approaching 'the environment' in Western societies, and within cultural geography much work has examined the meanings given to landscape and the ways those meanings affirm or contest particular forms of social relations. The previous sections of this chapter have already suggested that there is no single meaning given to 'the environment' because different ideas about gendered (and other) identities are reinforced or challenged by the different meanings the environment has been given. Feminist geographers have applied the same argument to landscape. Their work thus also contributes towards a feminist analysis of the environment. But it also pays particular attention to the ways in which environments are interpreted.

Landscape is a term which usually describes some kind of clearly delimited geography, very often a framed visual image of an environment. While people may talk of rural or urban landscapes, upland, lowland, wild or cultivated landscapes, some kind of organised scene is almost always implied. This scene is related to ways of seeing or picturing the world: a picturing which may be derived from our experience of being in the landscape, from written descriptions or from visual imagery. Landscape, then, refers to both material and imagined places. It is therefore also linked to issues of *representation* (see Box 6.3): how different peoples and places are shown in different media in specific times and places. What cultural geographers have emphasised about the term landscape is that different Western societies, or parts of these societies at different times and in different places, have shared views about ways of seeing, organising and feeling about the environment.

Box 6.3   Representation

Representation refers to the way in which interpretations are made of the world. It is a term which suggests that we do not perceive any aspect of the world in a naive and unmediated way, but that what we perceive is always re-presented to us through specific ways of making sense. Representations construct meaning about the world. They do so by using the codes, conventions and symbols of their specific historical, geographical and cultural contexts, and by referring to other familiar images. Western advertisements for holidays or cosmetics, for example, often contain images which remind audiences of countless other images of the 'rural', 'urban', 'exotic' or 'natural' and what they traditionally symbolise: simplicity, sophistication, sensuality. But images do not have single meanings. People can make sense of them in different ways and read or view images in opposition to the dominant culture. Richard Dyer writes:

> . . . we are all restricted by both the viewing and reading codes to which we have access (by virtue of where we are situated in the world and the social order) and by what representations there are for us to view and read. The prestige of high culture, the centralisation of mass cultural production, the literal poverty of marginal cultural production: these are aspects of the power relations of representation that put the weight of control over representation on the side of the rich, the white, the male, the heterosexual. Acknowledging the complexity of viewing/reading practices in relation to representation does not entail the claim that there is equality and freedom in the regime of representation. (Dyer, 1993: 2)

Dyer is commenting on the *politics of representation*: the ways in which representations both construct understandings of the world and inform its material organisation. The design of a public park, for example, makes assumptions about its use by different social groups, and this in turn depends upon cultural understandings of 'nature', respectability and appropriate conduct in public space. Representations do things, they work, they have effects and are thus material. Collective and individual identities, and experience of oppression or opportunity in social life, are often inseparable from representations of people, and their relationships with 'nature', environment and landscape. Dominant and oppositional claims to how social life should be organised, collective and individual identities, and ideas of 'nature', environment and landscape are thus *mutually constituted through representation*.

There are perhaps two ways in which landscape images can be described as gendered. The first is in their content. Landscape images depicting figures of men and women or symbols of supposedly masculine or feminine spaces often suggest certain ideas about gender. For example, images of English cottages and gardens at the turn of the twentieth century often showed women as mothers, suggesting this as women's primary role and duty. But probably the most profound way in which landscapes can be gendered is through the long history in Western society of describing the natural, and by association the rural, world as feminine. As the previous section commented, there is a long tradition in the West of understanding 'nature' as feminine, and femininity as close to 'nature'. Linked to this idea is the tradition of describing the female body as a terrain or landscape. Both of these ideas rest on an associated tradition which connects creative representation with men and the objects which give pleasure to a male viewer as feminine. This means that landscape can also be described as gendered in its very form. Not only what it shows, but how it shows it, has been described as masculinist by feminist geographers, and this is explained in the next sub-section. However, as the sub-section after shows, many women geographers and artists are claiming positions from which to look, interpret and enjoy images of landscape themselves. By doing so they disrupt the understanding of landscape that fixes them into automatic ways of seeing that are determined through gender.

Discussion of landscape must, however, also consider social relations other than gender. This section examines class, race and sexuality, and the following exercise asks you to think about other social relations and their gendering of landscape images.

ACTIVITY

Buying images of landscapes and bodies
The aim of this project is to investigate the images for and of women and men, that are marketed in the greeting card industry. It also asks you to think about how images of particular kinds of places are used to produce ideas of femininity and masculinity, sexuality, ageing, and conventions of physical beauty.

Greeting cards often use conventional traditions of representing particular kinds of places. They are also used to mark events in people's lives that are considered important (births, birthdays, coming of age, marriage, retirement, bereavement, etc.). Furthermore, different kinds of places and styles in cards aimed at different kinds of people (defined through age, gender and personal relation: for example, mothers, fathers, teenage women, baby boys, etc.) contain very strong messages about age and gender as well as sexuality. But they are also invested with meaning by those buying and giving them, written upon and given with particular intentions. This project on cards indicates the way in which personal meaning and mainstream imagery may intersect, how images from 'high culture' now widely circulate, how the cultural and economic are always interconnected, and how we can understand culture as a process of making meanings through practices as well as images.

- Visit card shops in your area and do a quick survey of card types and analysis of their content.
- What art historical styles are popular? (e.g. abstract, impressionist, black-and-white photography, pre-Raphaelite?)

- What kind of connotations do these styles and media have?
- Can you identify any major groups of geographical subjects? (e.g. domestic, rural, urban?)
- What activities are linked to men and to women in these images?
- What sort of men and women are depicted? (young or old, black or white, able-bodied or disabled?)
- Are particular figures in terms of gender, sexuality, race or age linked to particular places? (e.g. garden, jungle, race track?)
- Does the person for whom you are buying the card affect your choice? How?
- In cards with text, what kinds of cards are produced for sending to mothers, fathers, wives, husbands, teenage women or teenage men? Consider types of places, styles and colours.
- What kinds of romantic images are there?
- What makes some places seem erotic? (beach, garden, road, bedroom?)
- How does the road, through cars and motorbikes, seem to be about youthful sex?
- Are these images marketed for men or women?
- Does being on the road have different connotations for men and women?
- What does buying a card of a female or male nude mean for you?
- Are male nudes presented in the same way as female nudes?
- How is ageing represented?
- How do radical cards work with or against genres of visual imagery?

There is, then, the possibility for many different ways of seeing and using landscape images which maintain a critical approach as to how an understanding of ourselves and others is produced through representation of landscape. This does not mean that a landscape image will always mean the same thing. Women and men with diverse senses of their own identity (through gender but also class, race, sexuality and so on) may make images of landscape in quite new ways. Thus, this section aims to show how ideas about gender, class, race and sexuality are closely connected in images of landscape, but it also examines how both women and men have complex responses to landscapes and how ideas and images of landscapes can be used both to question and to support dominant meanings about the world. In doing so this section will raise questions about how feminist perspectives might help us understand the different and contested meanings of landscapes. Landscape images are always negotiated, and in this sense they have no final meaning. This suggestion that feminists should remember the diversity of landscape meanings places differences among women at the heart of feminist geographies of the environment.

Landscape as a 'way of seeing'

'Landscape' is not a simple thing. It is certainly not an inert object which can simply be observed, whether by geographers in the field, or by tourists in an art gallery, or by hill walkers from a vantage point. Rather, landscape is a term which makes sense of the environment in a particular way. To describe a certain perception of an environment as a landscape implies a particular interpretation of an environment. For example, landscapes tend to be rural.

Although they may contain human figures, these are usually subordinated to the surrounding trees, rocks, rivers and sky. And landscapes are most often structured as a kind of scenario laid out before their spectators – whether those spectators be geographers, tourists or hill walkers. Landscapes usually offer a panoramic view of an environment, which the gaze of their observer can sweep across. Several writers have argued that it is this specific combination of the visual and spatial organisation of an environment which is the defining quality of landscape. Landscapes are seen, and they are seen in a particular way, through a space which represents the environment as a territory stretching out before its spectator. For Denis Cosgrove, a cultural geographer who has often written about landscape, landscape should therefore be understood as a 'way of seeing' (1985: 46). It is not only a material object, but also a visual way of organising the perception of the environment. Writing with Steve Daniels in 1988, Cosgrove argued that 'a landscape is a cultural image, a pictorial way of representing, structuring or symbolising surroundings' (Cosgrove and Daniels, 1988: 1).

Work on the visual representation of landscapes is relatively recent in Geography. It is part of the growing interest in the production of diverse geographical knowledges about the world discussed in Chapter Two. Early work by cultural geographers raised the question of social power relations in connection with landscape by focusing on class relations. Whether written or painted, grown or built, these geographers argued that a landscape's meanings draw on the cultural codes of the society for which it was made. These codes are embedded in social power structures, and theorisation of the relationship between culture and society by these new cultural geographers has so far drawn on a humanist Marxist tradition which emphasises the importance of the values embedded in those cultural codes for the production and reproduction of class relations (Daniels, 1989). For Cosgrove in particular, landscape is a way of seeing the world which expresses only the values of the dominant ruling class. Cosgrove (1985) points out that landscape first emerged as a term in fifteenth and early sixteenth century Italy. He argues that it was bound up both with Renaissance theories of space and with the practical appropriation of space. He discusses the development during the Renaissance of a range of geometrical skills: the rediscovery of Euclidean geometry, the invention of the technique of three-dimensional perspective in 1435, and improvements in the methods of cartography, surveying and ordnance. But he argues that all these skills and techniques were put to the use of only one particular group in Renaissance society: the urban merchant class. This group were buying estates in the countryside and used these skills to map and survey them; and they also commissioned paintings of their land, and these paintings were the first to adopt the way of seeing to be called landscape. Through the use of geometrical perspective, these paintings of the Italian countryside represented the environment as a scenario laid out before the spectator. Perspective made this scenario seem 'natural', simply what was there. Yet their perspectival way of seeing the environment reproduced the particular relationship to the land of the landowner. They established a particular viewpoint for the spectator in their painting: a single, fixed point of the bourgeois individual. From this position, the spectator controlled the spatial organisation of a composition, and Cosgrove argues this was central to landscape images. In these canvases, through perspective, merchants could enjoy perspectival as well as material control over their land. And it was not only how their lands were visualised which affirmed their classed position, it was also what was shown, for landscape images rarely contained pictures of the workers who toiled on the land. In both form and content, landscapes were represented from the landowner's perspective. Cosgrove concluded that the landscape is seen and understood from the social and visual position of the landowner. Landscape is meaningful as a 'way of seeing' bound into class relations, and can be described as a 'visual ideology' in the sense that it represents only a partial worldview (Cosgrove, 1985: 47).

Cosgrove's argument was innovative in a number of ways. He focused on the visual as something which could be organised in different ways, and he carefully connected both the content and the organisation of landscape to the dynamics of class relations. However, his work paid very little attention to gender relations. Cosgrove (1985) did mention in passing that the bourgeois, landowning spectator of landscape was male, but he did not elaborate the point. Nonetheless, it is possible to argue that thinking about gender in relation to landscape both enriches and complicates Cosgrove's arguments. On the one hand, landscape is a particular way of seeing the environment, or, more specifically, a way of seeing 'nature'. And, historically, 'nature' has been gendered feminine in Western cultures. So a feminist critique of landscape could also consider the way in which the content of landscape is gendered feminine. On the other hand, the particular way of seeing landscape through the organised and rational gaze of geometrical perspective has some striking similarities to the distanced and objective mode of knowing the world which, as Chapter Two argued, has been described as masculine by many feminists. So a feminist critique of landscape could consider how the way of seeing landscape is gendered masculine. Both of these possibilities have been pursued by feminist geographers.

It is possible to suggest that this masculine gaze at 'nature' is a gaze at an object which is constituted as feminine. As the previous section argued, 'nature' has been gendered as feminine in Western discourses, as the still commonly heard phrases Mother Nature or Mother Earth testify. This feminisation of 'nature' has produced a particularly exploitive approach to the environment. The previous section showed how cultural ecofeminists in particular have used this argument to explain why the natural environment has been so degraded. The same argument can be made in relation to landscape.

If what a landscape image looks at can be interpreted as feminised, the look itself has been described as masculine by feminist geographers. The particular kind of look which sees landscape – a look at a scene laid out in orderly space some distance away – has been argued by many feminists to be a look which is masculine. They argue that dominant forms of masculinity in the West assume a rational and objective self, and a self which is positioned in relation to a world perceived as separate from them (Lloyd, 1984). This distance is achieved in part through the adoption of a certain distancing and controlling gaze at the world. This masculine gaze is also the gaze which Cosgrove described as the way of seeing landscape: it places the world at a distance from the observer, and represents it through the rational, scientific organisation of perspectival space.

Figure 6.1 Gilpin, *Guides to the Picturesque*, from William Gilpin's *Three Essays: On Picturesque Beauty, On Picturesque Travel, and On Sketching* (1792)

All these arguments can be demonstrated in the case of the way of representing landscape images known as the Picturesque, which was a particularly influential way of painting landscapes in eighteenth-century Europe and North America. It represented landscape in what was to become the standard for Western European landscape. It constructs a distance from what it depicts, and it now seems very hard to represent a standard landscape scene with large trees that frame a landscape composed of receding planes of topography and distant horizon, unless you stand back and are at least slightly elevated (see Figure 6.1).

Although this is simply one convention for representing places, it gained great authority as the proper way to depict landscape in England since the eighteenth century. Not only were places described in poetry and painting through this Picturesque convention, but gardens and estates were organised to help achieve this view from the houses of the aristocracy and gentry. This ability to see the landscape in certain ways was not neutral, however. Those that defined their good taste and social standing through their ability to recognise and enjoy picturesque landscapes did not extend this privilege or pleasure to others. Neither women in general, nor those who worked the land, nor those defined as racially inferior, were thought by upper-class men with private incomes to be able to see the landscape in this way. Their closeness to it, through either ideas of femininity, racial identity or work, meant that they were essentially lacking in taste and the distanced objectivity which defined the ability both to govern and to see the world in appropriate ways. Thus the organisation of vision in landscape was deeply tied to the definition of upper-class, white, male, political and economic privilege. More widely, the history of landscape representation in Britain and the West helped produce ideas of the identities both of those in powerful positions and of those subordinate to them: women, working-class, non-white racialised groups.

So far we have argued that landscape must be understood as a way of seeing which is both classed and gendered. However, it is also important to remember that other social relations are produced and reproduced by landscape images. One such relation has been implicit in the discussion so far: heterosexuality. The assumption that masculine scientists construct an Other which they may fear but which they also desire as feminine implies a heterosexual organisation of desire. Moreover, as Jane Gaines (1988) has argued, race must also be considered when the power relations of a particular, dominant way of seeing are being considered. Gaines points out that the notion that there is a powerful masculine gaze which can look at the world with impunity ignores differences among men. In particular, she points out that for many black men in the American South until very recently, to be caught looking at a white woman was to risk being lynched. Gaines is arguing that the powerful look is in racist societies a white one; black looks do not have the same authority. This is the argument currently being elaborated by a number of writers interested in white travellers in European colonies during the eighteenth and nineteenth centuries, as Chapter Two mentioned. Several of them point out that the intersection of race with class and gender is complicated, and that in colonial situations white women could look with almost as much authority as white men because of their privileged racialised position (Blunt and Rose, 1994). This argument is also a context in which to place the criticisms of ecofeminism as a Northern perspective which the previous section raised. Studies of landscape demonstrate particularly clearly that white women do not look at the environment simply as women, but as women from a particular cultural tradition (and very often with relatively high levels of material resources).

ACTIVITY

Consuming 'heritage' landscapes

Historic houses and gardens, whether rural or urban, attract tens of thousands of visitors each year in a great many different countries. Visitors are confronted with particular constructions of landscape and history in which certain things are included whilst others are silenced or excluded and yet visitors will make very different 'readings' of the place. If possible, spend a few hours at a nearby historic house or garden. Record your impressions and feelings and think about the questions below, bearing in mind how you might deal with attraction to and enjoyment of such landscapes and remain aware of the histories of class, gender, sexual, racial and other oppressions that were (and are) part of their production.

- How do you feel visiting historic houses and gardens?
- What elements of the landscape do you think contribute to your sense of enjoyment or unease?
- How is the history of the place presented?
- Does it match other images in film or period drama or brochures of historic houses and gardens?
- How is the house/garden itself represented in landscape images, textual, painted or photographed?
- Can you see any evidence of hidden histories? Think about the location of labourers' cottages, kitchens and servants' quarters.
- Are there any clues to the source of the wealth? (colonialism, military activity, plantation agriculture, imperial commerce?)
- How is a sense of nationalism conveyed?
- How is (and was) a sense of authority produced?
- What does the garden include or exclude? (types of vegetation, styles of gardening and architecture?)
- What kinds of family histories are celebrated?
- Does the family history in portraits and other memorabilia say anything about the power relations in the house or about acceptable gender and sexual relations?
- Do these meanings still affect you?
- What sort of groups seem to be visiting the house?
- Do you feel included or excluded?
- Is there any chance for expressing anger or disappointment or ambivalence?
- What do any graffiti or visitors' books say?
- Are there certain places where 'disrespectful' behaviour is more acceptable or contained? (toilets, rooms, parkland?)
- In what way is your behaviour constrained? (signs, wardens, other visitors?)
- Could the meaning of that place be changed for you and others by doing anything personally or openly resistant? (in your dress, speech, secret shared sign?)

- How does your buying of postcards or souvenirs or taking photographs change the meaning of the place for you?
- Is it possible to negotiate the pleasures which these places give you with an awareness of past and contemporary power relations produced by these places?

Feminist discussions of landscape emphasise the complexity of looking at landscape, then. They suggest that many kinds of social identities intersect in the representation of landscapes, and that such identities can be reproduced through both the content and the look at landscape.

Making feminist landscapes

As we have shown in the previous sections, whilst there may be dominant ways of looking at the environment, there are also resistant, oppositional and contested ways of seeing. The same argument must be applied to landscape images also. In other words, there are no 'real' or 'true' landscapes to discover beneath paint, text or vegetation but rather multiple, simultaneous, different and sometimes competing ways of seeing landscapes. Landscapes are not merely 'there' on the ground but are socially constructed within a complex and changing interplay of power relations, not least those between gender, class, race, sexual preference and other social differences. This sub-section is concerned with the instability and contradictions of landscape representations, and it will focus on the multiple ways in which landscape can be made and interpreted.

Many feminists have emphasised the ways in which different audiences make sense of the same images in different ways, in order to problematise the production of powerful, gendered knowledges about landscape. The meanings articulated in the production of a landscape image or built environment may not be the meanings interpreted by the users of the environment or the viewers of a landscape. The consumption of environments and landscapes by different audiences may be diverse and even conflictual. This is the emphasis of many accounts of consumption (Burgess, 1990), and it is an important argument in many feminist accounts of the politics of representation because it insists that there are always possibilities for diverse feminist readings of images. The following activity explores some of the different ways in which the paintings of Georgia O'Keeffe have been received by art critics.

ACTIVITY

Art critics' reactions to the landscapes of Georgia O'Keeffe
This exercise aims to explore a range of approaches to interpretation and illustrates the importance of context and specificity (the who, what, where and when) in evaluating interpretations of landscape, environment and 'nature' and the use of particular ideas of gender and sexuality within representation. It focuses on one image, *Flower Abstraction* (1924) by Georgia O'Keeffe (Figure 6.2), and is designed to encourage critical thinking about both your individual interpretation and the views that others offer of the meaning of this painting.

Spend a few minutes looking at and thinking about this painting. What does it bring to mind? If you like it, don't like it or feel indifferent about it, try to work out why. Jot down your thoughts about it.

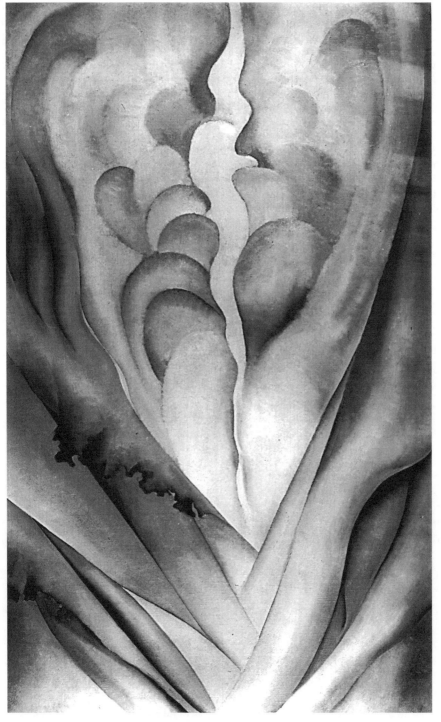

Figure 6.2   O'Keeffe, *Flower Abstraction* (1924)

It is generally thought that Georgia O'Keeffe used natural landscape and flowers in particular to represent women's bodies, but this has been interpreted and commented on in many different ways. The extracts of writing below do not agree on the meaning or significance of her work. They exemplify the way in which one image may be read or viewed in different ways. Although most make some kind of connection between the aesthetic value of her work and its social significance, they differ in their use of ideas of 'nature', landscape, biology, the body, gender difference and femininity. O'Keeffe's work has been read as celebrating women's closeness to 'nature', or as strategically representing the female body through images of flowers in order to provide images of the female body which differ from the male-dominated traditions of Western visual culture. Some comments on it seem to make assumptions about an audience of men sharing male heterosexual pleasure in the female body, while others risk reasserting oppressive ideas of women as close to 'nature' and imply an assumed reproductive/biological/genital basis to femininity and to gender difference. Thinking about the politics of these interpretations does not mean deciding on one 'true' or 'correct' response. Rather, each approach needs evaluating in terms of its context and politics.

Read each of the extracts below while thinking of these issues.

Source (i)

> The pure, now flaming, now icy colours of [Georgia O'Keeffe], reveal the woman polarizing herself, accepting fully the nature long denied, spiritualizing her sex. Her art is gloriously female. Her great painful and ecstatic climaxes make us at last to know something the man has always wanted to know. For here, in this painting, there is registered the manner of perception anchored in the constitution of the woman. The organs that differentiate the sex speak. Women, one would judge, always feel, when they feel strongly, through the womb. (Paul Rosenfeld, 1921)

Source (ii)

> First then, if the work [*Red and Yellow Cliffs*, 1940] resembles cliffs, it also at least in part establishes a relationship of resemblance to mounds and folds and furrows of flesh, and specifically to the human vulva, an association that the ambiguity of scale serves to keep in play. (By what I take to be an accident of circumstance, the resemblance is underscored by the presence within the same museum collection of Egon Schiele's *Beautiful Girl I Saw in a Dream* of 1911 ..., in which the spread labia of its subject are pictured in a manner strikingly similar to the way in which the central clefts are pictured in O'Keeffe's landscape. It may be acknowledged that similarity is a potentially distracting relationship. What we can at least say, however, is that for anyone who had the Schiele's in mind, the association of cleft with vulva in the O'Keeffe would be rendered virtually inescapable.) (Charles Harrison, 1994: 221–222)

Source (iii)

> Women artists have used the central cavity which defines them as women as the framework for an imagery which allows for the complete reversal of the way in which women are seen in the culture. That is: to be a woman is to be an object of contempt and the vagina, stamp of femaleness, is despised. The woman artist, seeing herself as loathed, takes that very mark of her otherness and by asserting it as the hallmark of her iconography, establishes a vehicle by which to state the truth and beauty of her identity. (Judy Chicago, 1975: 143–144)

Source (iv)

> In trying to develop a (woman's) visual language of desire, however, O'Keeffe was on her own. Her solutions to that problem were, admittedly, uneven: now crude and

obvious, now elegant and ingenious. She rejected from the first the dominant modes of picturing desire: she did not depict in a literal way the site of desire itself, the human body. Not only did she deny the (male) viewer the opportunity to look in a sexually predatory way at actual female anatomy (though critics proved remarkably inventive even so in their voyeuristic readings of her art's metaphorical content), but she also eschewed the easy but self-defeating task of inviting (female) viewers to gaze at the male body. Instead, O'Keeffe portrayed abstractly, but unmistakably, her experience of her own body, not what it looked like to others. The parts of the body she engaged were mainly invisible (and unrepresented) due to their interiority, but she offered viewers an ever-expanding catalogue of visual metaphors for those areas, and for the experience of space and penetrability generally.... O'Keeffe's abstract and highly sensual images of often labia-like folds, sometimes rendered in pastel shades, invoked associations not only with the body, but with skies and cloud formations, as well as with canyons and the anatomy of flowers. O'Keeffe had intense feelings about certain elements of nature, especially the open skies and spaces of the plains, and she found in natural configurations, large and small, homologies for the felt experience of the body. (Anna Chave, 1990: 118–119)

Some questions to ask:

- How do the extracts differ in the ways they connect O'Keeffe's art, femininity and nature together?
- How might it be useful sometimes to assert feminine difference in art and in the experience of natural environments or the body?
- If the image was made by a man would their claims work in the same way?
- How could each of the extracts be rewritten in ways which maintain or alternatively disrupt conventional views of masculinity and sexuality?
- Does the gender of the writer make a difference to what you think of the comments?
- How do the different extracts imply certain kinds of audiences? Did you have a sense of sharing or not sharing their response? What may this mean?
- Return to your notes on your initial thoughts. How have your ideas about the image been affected by what you have read and thought through?
- Those who produce images of the body or landscape cannot control who sees or reads them or how they are understood. What, then, is at stake and at risk in the reproduction of feminist erotic or body-centred imagery?
- Can it sometimes be useful to assert essentialist difference or use gender as the basis of critical evaluation of the politics of an image or readings of it?

Feminists have looked at landscapes in many ways, and feminist critiques and re-viewings of landscape are diverse. This sub-section focuses on just some of the tactics adopted by feminist geographers. Feminist geographers have looked for spaces in landscapes which allow women to articulate their complex identities; they have read images for new and feminist meanings; they have studied work which explicitly challenges dominant ways of looking at landscape. Both these tactics emphasise again and again the complexity of landscape, so that generalising about a feminist landscape becomes difficult if not impossible. Moreover, this complexity also suggests that feminist projects may also be able to construct alliances with other marginalised groups, so that gender itself becomes a problematic category with which to work.

One theme of feminist geographers' work on landscape is to search for another way of looking at landscape so that landscape does not become a way of seeing accessible only to the powerful. Much non-feminist cultural geography has tended to focus on Renaissance estates or large eighteenth-century landscape parks, and their (male) designers. Landscapes apparently considered too everyday and banal have been ignored. But many of those other landscapes have also, historically, been spaces through which women have expressed a relationship to landscape, and feminist geographers have argued that they therefore deserve theoretical and substantial enquiry. One such everyday landscape is that of the garden. Gardens, other than the grand landscape park, have been gendered as feminine. They are thus an environment in which women can construct their own landscapes. While this might be seen as constraining, Ford (1989) has argued that the garden can provide a space for the assertion of landscape expertise otherwise denied to women. She has studied in particular the botanical writings of Jane Claudius Loudon, who in the nineteenth century became an expert in garden design and whose work legitimated the landscape-making activities of other bourgeois women. The middle-class nineteenth-century villa was a separated private space with the wife/mother at the domestic centre, and the naturalisation of her nurturing abilities allowed her to control and make specific garden spaces such as flower-beds.

The placing of women in positions understood as closer to 'nature' – gardens, rural places – can be reinterpreted in other ways too. Helen Allingham produced many images like *Cottage at Chiddingfold* (1889) reproduced in Figure 6.3.

Allingham was a highly acclaimed professional painter whose late nineteenth- and early twentieth-century watercolours of country cottages and gardens fed into a strong preservationist movement which felt that 'traditional' architecture was being undermined. As a woman artist, Allingham's mobility within the rural landscape was more constrained than it was for her male counterparts, yet she took pleasure in looking at and making images of those landscapes. Such pleasure was enabled by her middle-class and professional position which allowed her leisure to travel and paint, to objectify the scenes and people she viewed. The figures in Allingham's landscapes are almost entirely working-class women and children, suggesting a mobilisation of acceptable codes of feminine representation for exhibition within scenes of a certain idealised Englishness – white, rural, southern, maternal, feminine and heterosexual – all of which immediately suggest the possibilities of other concurrent landscapes. Deborah Cherry (1993: 182–183), a feminist who makes 'interventions into art history', argues that 'it was this powerful combination of the rural idyll and domestic femininity that enabled the working-class woman at the cottage gate to signify social order.' Men are conspicuously absent in Allingham's images of country cottages, perhaps naturalising the domestic positions occupied by women and young children, or even making visible the link in women's and children's subordination. Alternatively, however, it is possible to re-read such 'blanks' in the paintings to produce new meanings. Some con-

Figure 6.3   Allingham, *Cottage at Chiddingfold* (1889)

temporaries did this. Some drew attention to the insanitary condition of many rural cottages. Indeed, in response to the cottage in *On Ide Hill*, 'Far from finding the scene picturesque, the local doctor berated the artist for painting a house that had more fever in it than any other in the parish' (Cherry, 1993: 181). Feminists now can also re-view these images. The 'blanks' might suggest female homosocial spaces whilst awaiting the return of men from work. The paintings, frozen in time and space, might even suggest men that never return, men who are permanently absent and consigned to the paintbox with the sweep of the brush. It is possible, then, to not only contextualise Allingham's work within a tradition of painting which objectified women and children, but also to renegotiate the meanings of the images in relation to present day concerns.

Other geographical studies are also arguing that garden spaces be viewed as complexly intersected by relations of gender and other social relations. If Ford's work focuses on gender and class differences, Morris (1994) explores gender and national identity in the context of the First World War. Her focus is on masculinities. Her work suggests that national identity intersects with gender, but that this occurs in complex ways. British male soldiers and Christian clergy on the Western Front, British male prisoners in a civilian prisoner-of-war camp near Berlin, and the British government through the Western Front military cemeteries all toiled to make gardens under adverse conditions. Whilst a great many soldiers and prisoners made or visited gardens in their attempts to cope with the appalling stresses of combat, the government was keen to evoke national sentiments by reproducing images of the ideal English garden in its propaganda. In this propaganda it was assumed that everyone in Britain, as well as the soldiers themselves, would be able to identify with an English garden: England (and not Scotland, Wales or Ireland) was the symbol of civilised values. The propagandists were often at pains to suggest the gentle yet tenacious masculine 'nature' of the 'English' soldier–gardener in contrast to the supposedly barbarous masculinity of the German. Morris's work suggests that diverse masculinities were at stake in these First World War gardens, as well as notions about 'Englishness'. She has also suggested how sexual preference can disrupt heterosexuality of landscape as a masculine engagement with a feminised 'nature'. In a study of *The Well of Loneliness*, written by Radclyffe Hall in 1928, she shows how the lesbian protagonist has troubled relations with not only her English country home of birth, but its landscape, gardens and 'nature'.

Earlier it was suggested that, historically, women and female bodies had been associated with 'nature' in Western cultures. Even more specifically, women have been associated with gardens and flowers within masculinist and heterosexist discourses about femininity, (male) desire, virginity and 'purity'. But women have always produced images of gardens and transformed material landscapes through gardening. It was not until the late nineteenth century that women earned an independent living as practising gardeners and garden designers in Western societies. Gertrude Jekyll (1843–1932), an eminent middle-class English garden designer and theorist, was one such woman. She is best known for her contribution to the English cottage style of garden, not unlike many of the garden borders to be found in Helen Allingham's watercolours. And amongst wide-ranging works are her lesser-known contributions to the style of plantings used to commemorate male 'sacrifice' in the British First World War cemeteries on the Western Front. Unfortunately, Jekyll suffered from failing eyesight even before she turned to garden design, and during the last two decades of her life as she grew more blind she had to work increasingly on paper, being unable to travel and view any prospective scene (Massingham, 1982: 121). Whilst Jekyll helped promote a style of gardening that could be easily associated with domestic femininity, she, like Allingham, disrupted any simple relation by being an independent working woman as well helping to produce potentially ambiguous connections between gardens and masculinities. The English garden is not quite as tranquilly calm as its usual representations suggested.

Women making landscapes of rurality: the example of Laura Ashley
Laura Ashley is an international clothing and furnishing company which took its name from the woman who started the company with her husband in 1953. Laura Ashley designs draw heavily on representations of rural landscape but these images intersect with class differences and ideas of race, gender and sexuality. Representations of the rural, of country living and ideas of 'nature' have been connected to the construction of idealised ideas of the past and nostalgia for forms of social relations based upon the class structure and gender relations of this past 'golden age'. This activity aims to get you thinking about how rural landscapes and ideas of 'nature' are evoked in clothing fashions and in home decoration in Britain. The example of Laura Ashley is focused on here but you could also look at how images of cities, rural landscapes and 'nature' in Britain and other places are used in other advertisements to sell the products.

Carefully look through Laura Ashley fashion and home decoration catalogues. Think about the complex relationships between 'nature', femininity, domesticity, nationalism, class, race and images of rurality and rural landscapes. Refer to the following questions.

- Do you think the catalogues draw on images of 'nature' and the rural to sell the products? If so, why? Give examples.
- What elements of the landscapes used in the pictures make you think of this?
- Look very carefully at the layout of the room, fabrics and furnishings. What styles are dominant and how would you describe them?
- What accessories are used to make the sets look 'authentic'?
- What are the names of the fabrics in the catalogues? How are they described in the catalogues?
- What kind of colours are used?
- Is the use of countryside imagery and 'nature' made explicit in the forewords to the catalogues?
- What do you think these images of rural landscape say about rurality?
- Do you think that a sense of Britishness is conveyed? If so, how is it displayed as being essentially English/British?
- How are women represented in the fashion catalogues?
- Are they in rural settings? Give examples.
- What are the models doing in the pictures?
- What kind of femininity are these women conveying to you?
- Do you think a particular type of femininity is represented in rural landscape imagery?
- What understanding of the 'roles' and 'nature' of women is being represented?
- What clientele do you think the company is aiming at?
- Do you think that the clothes are being aimed at women of a particular class or age? If so, why?
- What kind of clothes are being sold?

- Compare them to clothing found in other catalogues. Are they different? If so, why?
- What kind of image do you think that wearing these clothes presents?
- Who do you think buys Laura Ashley clothes and furnishings?
- Would you buy the clothes/furnishings? Why or why not?
- What values do you think the company are trying to instil in and through rural imagery?
- Do some research into the company. Look for books that have been written about Laura Ashley and her history (e.g. Sebba, 1990; A. Pratt, 1992). Compare these to what you thought about looking through the catalogues.

Feminist interpretations of landscape, whether they be the writing of academic interpretations, the depiction of landscape through other creative media, memories or visits, are not easily separated. We have already outlined some of the ways in which feminist geographers' different excursions into landscape studies are not merely interpretations but are also a making and re-making of landscapes. The following discussion examines the work of some vision-based artists and designers who directly challenge the exclusions and constraints of dominant ways of looking at landscapes.

Ingrid Pollard is a black woman photographer. She has produced *Pastoral Interludes* (Figure 6.4), a series of black-and-white prints showing black figures in English rural landscapes. The presence of a black person at once critiques the 'whiteness' of dominant representations of England's 'green and pleasant land' which is more usually shown as inhabited and worked only by white people, and there for the leisure only of white people. Pollard's work makes a black presence in the countryside visible. The image reproduced here

Figure 6.4   Pollard, *Pastoral Interludes*

perhaps points towards the fear of lone women, particularly black women, within English hill-walking country: '...feeling I don't belong. Walks through leafy glades with a baseball bat by my side...'. Pollard herself comments that the baseball bat is also a reference to the English countryside as the location of shooting (grouse and so on) as a means of culling and exterminating other species that are given no place in the countryside.

The potential for reading Pollard as a *feminist* photographer is further emphasised in her critical photographs of English tourist and heritage landscapes with the photo-essay, *Another View* (Pollard, 1993) (Figure 6.5). Here, a black woman's hands are shown writing on postcards of Dorset and Hastings upon which are superimposed images of Pollard herself as an uneasy reminder of the 'whiteness' of tourist and heritage advertising and merchandise. In the photograph shown in Figure 6.5 Pollard is claiming a different position from which to look at

Figure 6.5   Pollard, *Another View* (1993)

and enjoy English landscapes (albeit an uneasy pleasure); a right to be there and a right to be represented and make representations. She challenges, disrupts and complicates the notion of a generalisable set of shared ideas about England and the implicitly white and masculinised position from which it is usually viewed.

Pauline Cummins is an Irish artist who deals with issues of sexuality, motherhood and cultural identity in Ireland (Nash, 1996). Her installation *Inis t'Oirr/Aran Dance* (1985) was made and exhibited in the mid-1980s in southern Ireland when women's reproductive rights were being debated and women were attempting to negotiate personal identity with traditional Catholic and nationalist ideas of Irish femininity. The Aran of the title and of the knitting tradition refers to the Aran Islands off the coast of County Galway in the west of Ireland, which were the focus of intense national desire and anxiety about racial, moral, linguistic, spiritual and sexual purity in the early twentieth century. The installation thus relates to a set of representations which constructed the landscape or imaginative geography of the west of Ireland as a highly significant and contested national landscape. The video installation is a series of slowly changing images of a male torso and images of wool, knitting patterns and an Aran jumper, accompanied by the voice of the artist. The artist narrates her thoughts about knitting, story telling, landscape and the movement of her hands over the male body while images of the body first dressed in the jumper then naked slowly evolve on the screen. By linking the domestic to an autonomous and active female sexuality through the emblem of traditional Irish rural life, Cummins prompts a reconceptualisation of the meaning of traditional crafts, the domestic and the feminine. To imagine that women knitting in past and contemporary Ireland could and can be absorbed in fantasies of sexual pleasure is radically disruptive of traditional versions of Irish femininity. The installation thus takes the nationalist and often patriarchal symbolism of the west of Ireland landscape to produce a feminist statement which brings together attachment to these specific cultural tradititions and a radical sexual politics.

Although it has been argued that women and female bodies have traditionally been associated with flowers, rural landscapes and 'nature' in Western discourses, these representations, as we have attempted to show, are not fixed and are not only produced or given single meanings by white, Western, heterosexual privileged men. Indeed, while some male artists have made masculinist, even mysogynist, images that symbolically represent female bodies, other men have themselves disrupted these associations between bodies, flowers and 'nature' (although not always unproblematically). David Wojnarowicz died of AIDS-related illnesses in July 1992. With the onset of life-threatening illness many of his images, combining photography and paint, took on symbolic and often quite explicit references to death, life, sex and the human body. Some of his mixed media work are of large exotic flowers (see Figure 6.6) which make immediate references to the works of Georgia O'Keeffe examined in a previous activity.

Whereas O'Keeffe's flower paintings have been again and again linked to female sexuality and female genitalia, Wojnarowicz's flowers have been interpreted as phallic, as a masculine 'nature' – but one infused with male homosexual desire, a desire resisting any dominant universalism. As an apt concluding remark

Figure 6.6   Wojnarowicz, *I Feel a Vague Nausea*

suggesting the politicisation of recognising the relations between dominant and other multiple ways of seeing, Wojnarowicz has been quoted as saying:

> ...each public disclosure of a fragment of private reality serves as a dismantling tool against the illusion of ONE-TRIBE NATION: it lifts the curtains for a brief peek and reveals the possible existence of literally millions of tribes. (Quoted in Lippard, 1994: 23)

Wojnarowicz's emphasis on differences so multiple, so variable, that any singularity of identity is threatened is paralleled by the work of some feminist geographers, mentioned in Chapter Three, who question even the integrity of gender as a category of social identity. Critical accounts of processes of representation, especially those emphasising the diversity of social identity and the articulation of that identity in the practices of both representation and consumption, can lead to such questions. But so too can other forms of feminist critique. In relation to environmental feminisms, for example, Plumwood (1992) also goes a step beyond recognising diversity, and calls for a complete reconceptualisation of the distinction between 'nature' and humanity. She argues that the world needs a new vision which renegotiates both masculine and feminine identities in order to break prevailing nature/culture dualisms and especially to free the concept of the human (culture) from the connection with the masculine.

Feminist engagements with 'nature' and landscape, then, have echoed the diverse understandings of gender found in feminist geography more broadly. Work by feminist geographers discussed in this section has argued that there is an essentially feminine relationship to 'nature' and landscape; that landscape and 'nature' must be understood as intertwined with gender and other social differences; and that the complexity of 'nature' and landscape is such that gender may not always be a central analytical category.

SUMMARY

- Landscapes are a particular way of representing environments visually.
- Landscape images can articulate the complex intersection of a range of diverse social identities.
- This complexity is such that some women, in some circumstances and some places, can appropriate landscape imagery for themselves.
- Other women resist landscape imagery, seeking other ways to represent their relation to the environment.
- This complexity is also such that those occupying other marginalised positions have also contributed to the reworking of landscape representations, and this can place the centrality of gender as an analytical category into question.

6.5   Conclusions

This chapter has explored three related terms: environment, 'nature' and landscape. It has attempted to clarify the diverse meanings of these terms. The environment usually refers to our physical surroundings. The first section of the chapter demonstrated that the physical surrounding of the built environment can be described as gendered because ideas about gender are articulated in the design of the environment and negotiated in its use. In the case of the so-called natural environment, it is often assumed that there is some kind of total separation between our human lives and that natural surrounding. However, the second section argued that, just like the built environment, the natural environment is also interpreted and used in ways which are highly gendered. Moreover, also like the built environment, the natural environment does not articulate gender alone; environments also mobilise ideas about race, class, sexuality, and so on. The second section took this issue further and began to question differences, not only among women in their relation to the environment, but also among feminists. To what extent do Western feminists draw on Western ideas about cultural difference to sustain their own arguments about the need for ecological sensitivity? If the practices of the design of the built environment and the management of the so-called natural environment are structured by patriarchal and other power relations, to what extent are feminists complicit with at least some of those relations? For many feminists, this is a question of representation: how Western women represent women different from themselves. The third section explored the question of representation more carefully, by looking at ways in which Western observers, men and women, have seen 'nature' by seeing landscape. This section concluded by emphasising the complexity of representational practices, and arguing for the complexity of social identities involved in both representing and consuming images of 'nature'.

That this chapter should conclude by emphasising complexity is not surprising. Environment, 'nature' and landscape are extraordinarily complex terms, because in feminist work they bring together many things usually separated as opposites: the material and the cultural, the natural and the cultural, the local and the global, the production and consumption of meaning. And they do so in complicated ways. In making and remaking the meaning and practices associated with these notions, multiple forms of identity intersect, and such forms are complex, often contingent and very often contested. Indeed, many feminists feel that the most inspiring feminist theorising now happening is inspiring precisely because it explores such fractured sites of material meaning, and argue that through such complexity many of the most oppressive certainties which have for so long functioned to keep women, and other marginalised groups, in their place may now finally be breaking down. Donna Haraway (1991) is one such writer who revels in the disruption of some of the most fundamental assumptions upon which much Western thought, including feminist thinking, has been based. She sees at the present moment an opportunity to question many divisions, including that between the human and the non-human, cultural and natural: but also between different social groups. Instead of difference and division, she imagines a world of mixing and uncertainty, a cross-over world where new possibilities beckon and new ways of living may be possible. This is perhaps a surprisingly optimistic vision at this moment of ecological and social devastation; but then breakdown is not such a terrible scenario for those who have little stake in what is being broken. The conclusion to this book – Chapter Seven – returns to this question and considers how feminist geography is responding to recent theorisations of cross-over possibilities.

Further reading

Anderson, K. and Gayle, F. (eds). 1992. *Inventing Places*. Sydney: Longman Cheshire. This is a collection of essays which explores the range of social relations through which environments, 'nature' and landscapes are given meaning, including gender.

Braidotti, R., Charkiewicz, E., Hausler, S. and Wieringa, S. 1993. *Women, the Environment and Sustainable Development: Towards a Theoretical Synthesis*. London: Zed Books. Good for an overview of the debates.

Merchant, C. 1981. *The Death of Nature: Women, Ecology and the Scientific Revolution*. London: Wildwood House. This is a classic discussion of the construction of a masculinised modern science and a feminised 'nature'.

Rose, G. 1993. *Feminism and Geography: The Limits of Geographical Knowledge*. Cambridge: Polity. Chapter 5 discusses dominant ways of seeing landscape, and some feminist alternatives.

CHAPTER SEVEN

Conclusions

NICKY GREGSON, GILLIAN ROSE, JULIA CREAM
AND NINA LAURIE

7.1   Introduction

ACTIVITY

Return to the activity in Chapter One which asked you to define feminism and to the responses you made to the questions we asked there. Looking at each of these defini-tions, can you identify which traditions of feminism (or non-feminist thought) these definitions may be identified with? Why do you position these definitions thus? In the light of what you have read in this volume have your definitions of feminism altered? If they have, how have they changed, and why? If they haven't, why haven't they? What do you now consider to be the most important issues for feminist geographers to address?

In their conclusion to the first feminist geography text, *Geography and Gender*, the WGSG state that a focus on gender relations greatly improves geographical analyses, and that gender relations are central to understanding gender inequalities; and they make explicit their political commitment to changing gender relations. They also identify gender relations and their varia-tion between local areas as a key area for subsequent research in feminist geography and call for the abandonment of the systematic subdivisions – eco-nomic geography, urban geography, social geography and so on and so forth – which figure so centrally in defining geography. From our position, writing in the mid-1990s, such claims seem characterised by a remarkable degree of con-sensus and certainty. The product of a particular moment in both the history of feminist geography and in geography, these claims suggest both a common ground and a common agenda for feminist geography. In comparison, con-cluding this book is an immensely difficult task. As we have shown throughout the preceding chapters, feminist geography is characterised by such theoretical and methodological diversity and difference that it would be entirely inappropriate, not to say impossible, to offer here a straightforward summary of feminist geography and its future. Instead, feminist geography in the 1990s is more appropriately thought of as feminist geographies, and the only prediction which would seem safe to make is that over the next few years these feminist geographies will continue to produce different feminist knowledges, different feminist interpretations and more and more diversity. In the remainder of this chapter, therefore, we highlight for your consideration some of those areas, questions and issues which we feel are likely to become central to future debate between feminist geographers. Many of these have been prefigured in the earlier chapters of this book, and indeed we hope that some might already have begun to occur to you. They are:

- Differences between women: differences of class, race, sexuality, able-bodiedness, age, location...
- Thinking about differences and strategies for dealing with difference
- Rethinking dominant definitions of space, place, landscape, environment and nature
- Writing difference and writing differently

Below, we explore some of these issues further. As elsewhere, we include a number of activities to help you think them through.

7.2   Decentring gendered geographies, rethinking gender geographies

The key concept around which this book has been organised is gender. However, as the chapters here demonstrate unequivocally, gender means many things to different individuals. Such differences are potentially difficult to handle. They could, for example, be seen as divisive and destructive; they could even be interpreted as undermining of the project of feminist geography. We, however, would argue differently: the differences among feminist geographers over the definitions of gender *do not* undermine feminist geography. Rather this book suggests that such differences can be put to advantage: gender is at the end of the day a category, something which we use to help us classify, analyse and interpret the world. And the different ways in which we put gender to work are productive, as well as highly constructive. They show, in short, how meanings and social relations are both constructed and contested; how our interpretations are produced through difference, and how feminist geographies themselves are made. However, and in contrast to the energy which feminist geographers have invested in thinking about gender, one of the most striking things to emerge in the course of writing this book has been the relative absence of debate about space and place. Chapter Five, for example, demonstrated that feminist geographers' notions and uses of space and place are frequently structured by dichotomous thinking: although the boundaries between home and work and formal and informal are seen as fluid, as contestable and as highly problematic, much of feminist geography has retained these dichotomies as a conceptual framework. The most commonly used strategy therefore within feminist geography has been to work from within existing ways of thinking about space to expose the people, things and experiences which they include and exclude. This type of work is likely to continue to characterise much of feminist geography in the immediate future. However, a tendency which is also likely to become increasingly prominent is one which tries to break with boundaries and bounded spaces, which tries to think about space in non-dichotomous ways. One such approach is to think in terms of what Gillian Rose (1993) refers to as paradoxical space. Thinking this way about space means working with the idea of being simultaneously inside and outside, occupying centre and margin – positionings which are frequently paradoxical according to dominant definitions of space and place. One such instance of paradoxical space cited by Rose is the work of Patricia Hill Collins (1990) on black women working as domestic workers in white homes. Being on close terms with the children in these families means that in many senses black women are insiders. But, they are also, through race and employment, made to know that they do not fully belong in these homes. The position which they occupy is one which is simultaneously present but absent, the outsider within.

ACTIVITY

Read Chapter 7 of Gillian Rose's book, *Feminism and Geography*. To what extent do you think this succeeds in breaking with dichotomous thinking?

Realising these points about gender and space, and the divergence in the extent to which we as feminist geographers have thought things through, may be hard to accept (see, for instance, *Antipode*, 1992). But at this point perhaps we need to remind ourselves of some of the oldest feminist lessons, namely that those issues which are hardest to face are often those which are nearest to home. Traditionally in feminist research this has meant paying greater attention to the personal, the bodily. And in terms of thinking differently about gender this is certainly one direction which some feminist geographers are pursuing. However, another way for feminist geographers to think closer to home is in terms of disciplinary homes, and the extent to which these need to be re-examined. In what follows we explore both of these issues a little further.

7.3   Embodying geographies

One of the most important aspects of the 'personal' to which feminists have paid great attention is the bodily. This particular material space has recently been suggested as the location for gendered analyses of space (Smith, 1993: 102), although as Chapters Five and Six here have amply demonstrated, spaces, places, landscapes and environments are gendered in some way in all their articulations.

The bodily is, however, also now being addressed by feminist geographers, and this is having some important implications for the key concept around which this text has been organised: gender. One of the most persistent themes to have emerged at various junctures throughout these chapters has been the question marks which increasingly are being raised over the centrality of the concept gender to feminist geographers. When *Geography and Gender* was

written, such was the centrality of gender to feminist geography that the concept figured not just in the title but as the category around which feminist
geographers' interests were seen to (and assumed to) coalesce. Now, however,
things are more uncertain, more provisional. For example, Chapter Three of
this volume outlined the multiple ways in which gender is thought through by
feminist geographers, concluding with a position which argues for the decentring of gender in feminist analyses, and similar arguments are articulated in
Chapters Five and Six in relation to the work of feminist geographers on
space, place and landscape. Such a position sits uneasily with those held by
many other feminist geographers, who – as we have seen – take the view that
gender (however this be thought of) sits at the centre of feminist analysis. In
this respect, then, feminist geography echoes precisely the dilemmas currently
being articulated in various other spheres of feminist scholarship. Fuelled by
recent developments within social and cultural theory, feminists are divided in
their response to the question of whether gender represents a point beyond
which the theorising has to stop. Feminist theorists such as Susan Bordo,
Nancy Hartsock and Sandra Harding are just some of those for whom gender
continues to constitute a 'significant difference', if not *the* significant difference (Haraway, 1991); gender for these writers is a theoretical category so
important to understanding the composition of social life and its material and
cultural effects that it should not be dispensed with. And in continuing to
centre gender in their analyses, many feminist geographers are clearly aligning
themselves with this tendency within contemporary feminist writing. Other
feminist writers, though, do not see things in the same way. They maintain
that, rather than being the significant difference, gender – or at least thinking
about this (as the significant difference) – represents much of the problem.
Gender, they argue, needs not just decentring but to be fundamentally
rethought. Foremost amongst those arguing in such a way is feminist theorist
Judith Butler, and it is to some of her arguments that we turn here.

In Chapter Three, when we posed the question 'what is gender?' (Section
3.2), we answered this in a way which accorded with the answer which most
feminist geographers (and indeed most feminists) gave to this question, at least
until very recently. This answer argues that gender is grounded in the binary distinction made between the male and female sex, and maintains that gender itself
represents the social constructions placed upon male and female, the attributes
and ways of behaviour which construct us as men and women, masculine and
feminine. This, to be fair, is probably the way in which the majority of feminist
geographers continue to think about gender even now. However, it is not the
only way in which it is possible to do so, and an alternative is suggested by
those who find the arguments of Judith Butler persuasive.

One of the central arguments made by Judith Butler in her book *Gender
Trouble* (1990) is that, rather than gender being a social construction based
on the binary opposition of the male and female sex, it is the two sexes themselves which are social constructions. Furthermore, Butler argues that
thinking about sex in terms of two 'natural' sexes is itself grounded in our
belief that the categories man and woman (i.e. gender) are defined in relation

to separate, distinctive bodies. The following summary expands on and illustrates these ideas.

What is gender? Gender and sex as social constructions

Rather than 'what is gender?', 'what is sex?' is perhaps a better question to ask. Sex is what makes our bodies male or female, but what is sex? What is it made up of? Is it made up of genes, genitals or hormones? Is a woman without a womb still female? Is a woman injecting herself with testosterone still a woman? Is a man carrying the chromosomes XXY still male? Perhaps sex cannot be reduced to biology. Perhaps there is no such thing as 'natural' sex?

The notion that we all fit into either a male or female body is just that, notional. It cannot be sustained, either over time or space. Simply put, there is nothing 'natural' about 'male' and 'female' bodies. There is nothing natural about everyone being forced into one sex or the other. Rather, our belief in the existence of two, and only two, sexes is structured by our ideas about gender. What this means is that our understanding of gender (man and woman) is not determined by sex (male and female) but that our understandings of sex itself are dictated by an understanding that man and woman should inhabit distinct and separate bodies. So, sex does not make gender; gender makes sex (Cream, 1995).

However, there are many bodies which do not fit the gender system which we have created. One of the clearest illustrations of the way in which the two-sex model can be undermined, as well as of the extents to which we go to preserve the myth of gender, is the case of intersexed infants. In research which has looked at the case management of children in the USA who are born with genitals which are neither clearly male nor female, Suzanne Kessler (1990) has shown how we literally manufacture sex and gender. Kessler found that even when doctors are faced by unequivocal evidence that the binary notion of two, and only two, sexes is a myth, they persist in making a child clearly born with both sexes *either* male *or* female. Their bodies are literally reconstructed, both hormonally and surgically. Such action highlights the fact that it is culture, and not simply biology, which dictates sex. Kessler also tells of how difficult it is to deal with a child that is not easily sexed. She notes that parents are often told to give their child a sexually ambiguous name, to buy the doctors time to decide which sex the child will be. She notes how doctors offer delaying strategies, ways of warding off the persistent question: 'is it a boy or a girl?'.

It is not, however, only those who lie at the margins who have to work hard at sustaining the myths of gender, but those at the centre, 'natural' men and women. As Suzanne Moore reminds us:

> The huge performance in order to display femininity – the waxing and workouts, the dieting, the skincare, the make-up, the accessorising and the clothes that goes into producing an acceptable face of womanhood over a lifetime, costs more than a sex-change operation.

She makes clear that it is not just those living on the edges of gender who have to work hard at maintaining its illusion of masculinity and femininity, but those who live at its centre as well, 'natural' men and women. We continue to perpetuate the myth, and it costs us. We pay for it in terms of cosmetic surgery and eating disorders, and also in terms of violence and discrimination. It is time to give up the myth of two, and only two, sexes.

ACTIVITY

Which answer to the 'what is gender' question do you find most persuasive? Why?
Construct a series of questions which you think might figure on various research agendas within feminist geography over the next few years. Take both answers to the 'what is gender' question as your two frames of reference here.

Bodily geographies

So, what does all this suggest in terms of future debate amongst feminist geographers? For those influenced by the arguments outlined immediately above, the existence of gender *and* sex as historically and geographically variable categories means that there is a need to think of different ways of understanding and talking about our bodies, our sex and our gender. These feminist geographers wish to produce feminist geographies which break away from the notion of biological natural sex, male and female sexed bodies, and there seems little doubt that their interests and concerns will focus increasingly on embodied performance; on bodily display; and on the acting out of multiple, fluid sexual identities. Recent studies in this vein include the essay by Bell *et al.* (1994) on gay skinheads and lipstick lesbians, and Lewis and Pile's (1996) paper on the Rio Carnival. Lewis and Pile suggest what effects the emphasis on bodily performance may have on understandings of space and place. Instead of understanding space and place as a pre-existing location in which performances take place, they argue that performances themselves constitute spaces and places in ways which are at once material and cultural. In this way, spaces and places become dynamic and shifting articulations of social relations of meaning.

The processes through which bodily performances are constituted through particular spaces are beginning to be explored in more detail by some feminist geographers concerned with the processes of subjectivity. The ways in which bodies themselves are imagined as spaces, and the spaces which they are imagined as inhabiting, are being examined by some feminist geographers in relation to a range of subjective, emotional and psychic processes. For example, this book has already noted that the space through which most Geography observes the world – very often a space imagined as transparent and completely knowable – is one symptom of the masculinism of the discipline. But geographers such as Rosalyn Deutsche (1991) and Gillian Rose (1993, 1995a) have suggested that while such a space constitutes the rationality of the masculine observer, it is also a space riddled with fantasies, pleasures, fears and horrors. Such emotions are often projected onto feminised spaces and/or bodies and expelled from masculinised bodies and/or spaces. Both Deutsche and Rose argue that one strategy of critique is then to insist on the subjective construction of material spaces, and that this strategy then also marks the embodiment of the production of geographical knowledges.

The body, then, is firmly on the agenda in certain quarters of feminist geography in the 1990s, and it is placed there by virtue of seeking an alternative answer to the 'what is gender' question to that which has dominated feminist geography to date.

7.4   Rethinking the material and the imagined

This consideration of the bodily by feminist geographers is related to another large issue with which many feminist geographers are engaging: an effort to connect the material and the cultural. As the previous section argued, work on performing gender suggests that performances are at once material (since they are embodied) and cultural (since those performances become meaningful in relation to cultural values, ideas and conventions). Efforts to think about other aspects of geography as both material and cultural are also under way in feminist geography.

One way of engaging with the natural and the cultural at once is suggested in Chapter Six. Parts of that chapter can be read as sounding a note of caution about the claim that everything is a cultural construction, that everything is an effect of representation. In particular, the discussion in Section 6.3 suggested that the environment is not entirely an effect of representations. Although interpretations of the material or 'natural' environment do vary greatly, the environment itself is not entirely reducible to those interpretations. While the effects of finite resources, pollution and deforestation may not be entirely known, or even knowable or predictable, they may nonetheless impose some limits on human action. This has led Donna Haraway (1991) to describe the environment as an *actant*. She argues that the environment is something which acts in ways precisely as complex and contradictory as human activity, and should therefore be understood as being as active as the human. She also argues that the human and the cultural are so completely intertwined that they should be considered together by using the same category for both: actant. This is perhaps an especially important move for feminist geographers to consider, since the discipline of Geography is so clearly divided into physical and human geography, a divide which reiterates the idea of the natural and the human as being in some profound sense different from each other. It makes the notion of a feminist physical geography quite possible, except that the label 'physical' would no longer make sense!

Another important area in which feminist geographers are displacing the distinction between the material and the cultural is in their engagements with what are often called 'imagined geographies'. These are geographies constructed through cultural maps of meaning. They have been a particularly important part of the new cultural geography (see Chapter One, Section 1.2), but many feminists have argued that non-feminist geographers have understood imagined geographies in ways which do not pay enough attention to the materiality of social life. For some feminist geographers, this is because, for example, the embodiedness of imagined geographies is not specified carefully enough so that the person interpreting and producing an imagined geography ends up occupying that distanced and unsituated analytical position which so many feminists have argued is constitutive of dominant masculinities (Deutsche, 1991). Other feminists have suggested that to ignore the materiality of imagined geographies neglects the unequal distribution of material resources and in particular the way that class structures relationships to imagined geographies, and differentiates therefore between gendered relationships to them (Gregson, 1993).

Several feminist geographers have therefore called for 'imagined geographies' to be understood not simply as imagined, not simply as cultural, but also as materially grounded. While acknowledging the importance of geographies that are imagined, they insist that such imaginings always take place through a highly differentiated set of power relations, and that feminists should consider the material context of these when engaging with imagined geographies (Bondi, 1993; G. Pratt, 1992). Examples of this argument examine diverse kinds of imagined spaces. Robyn Dowling (1993), for example, describes the department store as a particular space constituted through building design, commodity display and the performances of diverse femininities. She interprets the complexity of this space by considering the relations of power which structure its meanings by establishing boundaries and surveilling performances; and she points out the differences between the way women were expected to perform as consumers and as shopworkers. Other feminist geographers have examined the construction of the nation as an imagined geography. Sarah Radcliffe (1996), for example, discusses the gendering of Ecuador as a nation, and argues that there are three elements to this process: domestic space becomes a trope for national space; women's bodies symbolise the limits of the national space; and racialised differences among women are represented as obstacles to national integration. Radcliffe's essay pays a good deal of attention to the way classed and racialised differences structure the feminisation of modern Ecuador, and she concludes by suggesting that such differences obscure the ways in which all women are positioned in quite particular ways by this imagined geography. Radcliffe, then, is both working with the notion of gender as an analytical category but also engaging with its profound differentiation. Her efforts to work both with and against the analytical categories to hand provides an excellent illustration of the sort of complex approach to gender outlined in Section 7.1.

Catherine Nash (1994) also addresses the feminisation of the nation-state; but she pays a good deal of attention to the work of feminist artists concerned to reimagine contemporary Ireland. In the Irish context, traditional ways of living in rural areas are particularly resonant images of national identity, and Nash points out that many such images focus on rural women as the epitome of Irish virtues, and often represent such women as part of the Irish landscape. Feminist artists have both parodied such constructions of femininity – making, for example, drawings of female bodies as geomorphological and surveyed landscapes – and displaced them, rendering the natural landscape itself as a construction. Nash thus emphasises the diverse critical strategies feminists may deploy in order to subvert dominant imagined geographies of the nation, strategies which problematise the distinction between the natural and the cultural by playing with the simultaneously material and cultural gendering of both.

7.5   Feminist geography–human geography

The relationship between feminist geography and human geography in general is one which, again, has been touched on at various junctures in this book, notably in Chapter Two, where we examined various ways of approaching the history of feminist geography and their connections with the history of geography literature; in Chapter Three, where we explored, amongst other things, the development of geographies of women as a counter to the masculine assumptions prevalent within Geography's subject matter; and in Chapter Six where we looked at feminist geographers' work on landscape and contrasted this with some of the work of cultural geographers and those working on the 'natural' and built environment. As all of these instances show, the relationship is one which can be characterised by tension, and in some cases (see, for example, Sections 2.2 and 2.4) a not inconsiderable amount of barely disguised hostility. That this is so is perhaps not surprising. Feminist geography emerged as a critique of the content of human geography. Indeed, one of the first papers to be written in feminist geography took geography to task for omitting 'half the human' from its consideration (Monk and Hanson, 1982). And since then, feminist geographers have gone on to critique further geographical knowledge itself (G. Rose, 1993), as well as the methods and research practice deployed within much of human geography (Nast, 1994; Katz, 1994; Kobayashi, 1994; England, 1994; see also Chapter Four). In offering this account of the relationship between human geography and feminist geography, it is possible to apply some of the arguments about space and place already made in this chapter. Thus the geography of this relation is at once material and cultural, in the sense that the materiality of institutional power structures interweaves with the theoretical maps of meaning used by feminist geographers.

Feminist geography could be said to occupy the borderlands or the margins of human geography, rather than the centre ground; even to be positioned, through its focus on critique, in a (self-imposed) land of exile. And yet, there is a sense in which many feminist geographers clearly do wish to belong more centrally within human geography. Many review articles and overviews, for example, seek evidence for feminist geography transforming the content of human geography itself, and are disappointed by their failure to find much evidence for this (see, for instance, McDowell, 1993a, 1993b) or angered by what they see as feminist geography's continued marginality to the projects of human geography (Christopherson, 1989). We, however, are of the view that to position feminist geography entirely on the margins of contemporary human geography is misplaced. Rather, for this collective, feminist geographies occupy the centre and the margins simultaneously: another example of paradoxical space. Constructed in response to many of the ideas which dominate human geography, and challenging of these ideas, feminist geographies depend on the existence of these dominant ideas for their ability to resist them, to challenge them and to construct alternatives. Where we differ, however, is in the extent to which we as individuals see an opposition between human geography and feminist geography, in the extent to which we see lines of tension between occupying both centre and margin. For at least some of those who have been involved in writing this book, there is a sense in which the almost taken-for-granted assumption within feminist geography of the opposition of mainstream human geography to feminist geography is beginning to be questioned. These members of the writing team would point to certain areas of human geography, notably those most strongly influenced by recent developments in social theory and cultural studies – areas of human geography in which debates over identities, representation and culture are of critical importance – and argue that these have much in common with, and are of considerable interest to, those versions of feminist geography which emphasise similar issues. For them, to talk of an opposition between particular visions of feminist geography and such versions of human geography would be misplaced in the extreme. And, as individuals active within and promoting of such visions of human geography, including feminist geographers, are promoted to positions of power and authority within the discipline, then, for some feminist geographers, it is more appropriate to talk of a growing rapprochement between areas of human geography and areas of feminist geography. Others of the writing team, however, are less convinced. They maintain that these key areas, which some see as indicative of similarity and rapprochement, are producing an alternative form of masculine geographical knowledge, albeit non-hegemonic within human geography itself, rather than feminist geographies. Ultimately, then, we cannot agree on how we interpret the current relationship between feminist geography and human geography.

ACTIVITY

Discuss amongst yourselves how you see the relationship between human geography and feminist geographies. To help get you started here you might like to recap how you have felt at various stages about doing a course in feminist geography. Start by thinking about your feelings prior to commencing your course. Assuming that your course is optional, how did those not taking the course respond to your decision? Then think about how your feelings and opinions may or may not have changed as you've worked your way through the course. Does this suggest anything to you about the relationship between human geography and feminist geography? Think here about institutions and individuals, as well as about the knowledges produced by both human geography and feminist geography.

Bibliography

AGARWAL, B. 1986. *Cold Hearths and Barren Slopes: the Woodfuel Crisis in the Third World.* London: Zed Books.

AGARWAL, B. 1992. The gender and environment debate: lessons from India. *Feminist Studies,* 18(1), 119–158.

AGGER, I. 1992. *The Blue Room: Trauma and Testimony among Refugee Women. A Psycho-Social Exploration.* London: Zed Press.

ALLEN, J. 1988. Fragmented firms, disorganised labour. In Allen, J. and Massey, D. (eds), *The Economy in Question.* London: Sage, pp. 184–228.

ANDERSON, B. 1983. *Imagined Communities: Reflections on the Origin and Spread of Nationalism.* London: Verso.

ANDERSON, J., BROOK, C. AND COCHRANE, A. (eds) 1995. *Global World? Re-ordering Political Space.* Open University and Oxford University Press.

ANKER, R. D. 1983. Female labour force participation in developing countries: a critique of current definitions and data collection methods. *International Labour Review,* 122(6), 709–723.

ARCHAMBAULT, A. 1993. A critique of ecofeminism, *Canadian Woman Studies,* 13(3) 19–22.

BABB, F. 1990. Women and work in Latin America. *Latin American Research Review,* XXV(2), 236–247.

BAINBRIDGE, B. 1993. Her own woman. *Art Review,* 45 (June), 56–57.

BARKER, I. 1994. *The Strategic Silence: Gender and Economic Policy.* London: Zed Books.

BARRETT, H. AND BROWNE, A. 1993. The impact of labour-saving devices on the lives of rural African women: grain mills in the Gambia. In Momsen, J.H. and Kinnaird, V. (eds), *Different Places, Different Voices: Gender and Development in Africa, Asia and Latin America.* London: Routledge, pp. 52–62.

BARRIOS DE CHUNGARA, D. 1978. *Let Me Speak: Testimony of Domitila, a Woman of the Bolivian Mines.* London: Stage One Press.

BAUD, I. 1992. *Forms of Production and Women's Labour: Gender Aspects of Industrialisation in India and Mexico.* New Delhi and London: Sage.

BELL, D. AND VALENTINE, G. (eds), 1995. *Mapping Desire: Geographies of Sexualities.* London: Routledge.

BELL, D., BINNEY, J., CREAM, J. AND VALENTINE, G. 1994. All hyped up and no place to go. *Gender, Place and Culture,* 1 (1), 31–48.

BENERÍA, L. 1981. Conceptualising the labour force: the underestimation of women's economic activities. *Journal of Development Studies*, 17(3), 10–28.

BENERÍA, L. AND ROLDAN, M. 1987. *The Crossroads of Class and Gender. Industrial Homework, Subcontracting and Household Dynamics in Mexico City.* Chicago: University of Chicago Press.

BERGEN, R.K. 1993. Interviewing survivors of marital rape: doing feminist research on sensitive topics. In Renzetti, C.M. and Lee, R.M. (eds), *Researching Sensitive Topics*. London: Sage, pp. 197–211.

BERGER, M. AND BUVINIC, M. 1989. *Women's Ventures Assistance to the Informal Sector in Latin America*. West Hartford, Connecticut: Kumarian Press.

BIEHL, J. 1991. *Rethinking Ecofeminist Politics*. Boston: South End Press.

BLAUT, J. 1961. Space and process. *The Professional Geographer*, 13(1), 1–7.

BLUNT, A. 1994. Reading authorship and authority: reading Mary Kingsley's landscape descriptions. In Blunt, A. and Rose, G. (eds), *Writing Women and Space: Colonial and Postcolonial Geographies*. London: Guildford Press.

BLUNT, A. AND ROSE, G. (eds), 1994. *Writing Women and Space: Colonial and Postcolonial Geographies*. London: Guildford Press.

BONDI, L. 1991. Gender divisions and gentrification: a critique. *Transactions of the Insitute of British Geographers*, 16(2), 190–198.

BONDI, L. 1992a. Gender and dichotomy. *Progress in Human Geography*, 16(1), 98–104.

BONDI, L. 1992b. Gender symbols and urban landscapes, *Progress in Human Geography*, 16(2), 157–170.

BONDI, L. 1993. Locating identity politics. In Keith, M. and Pile, S. (eds), *Place and the Politics of Identity*. London: Routledge, pp. 84–101.

BOWLBY, S. 1992. Feminist geography and the changing curriculum. *Geography*, 77, 349–360.

BOWLBY, S. *et al.* 1989. The geography of gender. In Peet, R. and Thrift, N. (eds), *New Models in Geography*, Volume Two. London: Unwin Hyman, pp. 157–175.

BRAIDOTTI, R., CHARKIEWICZ, E., HAUSLER, S. AND WIERINGA, S. 1993. *Women, the Environment and Sustainable Development: Towards a Theoretical Synthesis*. London: Zed Books.

BRAND, D. 1990. Bread out of stone. In Scheier, L., Sheard, S. and Watchel, E. (eds), *Language in Her Eye: Views on Writing and Gender by Canadian Women Writing in English*. Toronto: Coach House Press, pp. 45–53.

BROWN, P. 1991. How refugees survive. *New Scientist*, 3 August, 21–26.

BROWN, W. 1988. *Manhood and Politics*. Totowa, NJ: Rowman and Littlefield.

BRUNETTA, G. 1995. Women immigrants in Italy. Paper presented at the International Conference of Population Geography, Dundee, UK, September 1995.

BRUNO, M. 1995. The second love of worker bees: gender, employment and social change in contemporary Moscow. Paper presented at the Institute of British Geographers' Annual Conference, January 1995.

BUCKINGHAM-HATFIELD, S. 1994. Gendering 21. *Town and Country Planning*, July/Aug., 210–211.

BUNGE, W. 1962. *Theoretical Geography*. Lund: Gleerup.

BURGESS, J. 1990. The production and consumption of environmental meanings in the mass media. *Transactions of the Institute of British Geographers*, 15(2), 139–161.

BUTLER, J. 1990. *Gender Trouble*. London: Routledge.

BÜTOW, B. *et al.* 1992. *Frauen in Sachsen: zwischen Betroffenheit und Hoffnung*. Leipzig: Rosa-Luxemburg-Verein.

CHANEY, E. AND BUNSTER, X. 1989. *Sellers and Servants Working Women in Lima, Peru.* Granby, Massachusetts: Bergin and Gavey.

CHANEY, E. AND GARCÍA CASTRO, M. 1989. *Muchachas No More. Household Workers in Latin America and the Caribbean.* Philadelphia: Temple University Press.

CHAVE, A. C. 1990. O'Keeffe and the masculine gaze. *Art in America,* 78(1), Jan., 114–125.

CHERRY, D. 1993. *Painting Women: Victorian Women Artists.* London: Routledge.

CHICAGO, J. 1975. *Through the Flower: My Struggle as a Woman Artist,* 1982 edition. London: The Women's Press.

CHOUNIARD, V. AND GRANT, A. 1995. On being not even anywhere near 'The Project': revolutionary ways of putting ourselves in the pictures. *Antipode,* 27(2), 137–166.

CHRISTOPHERSON, S. 1989. On being outside 'the project'. *Antipode,* 21(1), 83–89.

CLOKE, P., PHILO, C. AND SADLER, D. 1989. *Approaches to Human Geography.* London: Paul Chapman.

CORRIN, C. (ed.) 1992. *Superwomen and the Double Burden.* London: Scarlet Press.

COSGROVE, D. 1985. Prospect, perspective and the evolution of the landscape idea. *Transactions of the Institute of British Geographers,* 10(1), 45–62.

COSGROVE, D. AND DANIELS, S. (eds), 1988. *The Iconography of Landscape.* Cambridge: Cambridge University Press.

CREAM, J. 1993. Child sexual abuse and the symbolic geographies of Cleveland. *Environment and Planning D: Society and Space,* 11, 231–246.

CREAM, J. 1995. Re-solving riddles: the sexed body. In Bell, D. and Valentine, G. (eds), *Mapping Desire: Geographies of Sexualities.* London: Routledge, pp. 31–40.

DANIELS, S. 1989. Marxism, culture and the duplicity of landscape. In Peet, R. and Thrift, N. (eds), *New Models in Geography* (Volume II). London: Unwin Hyman, pp. 196–220.

DANKLEMAN, I. AND DAVIDSON, J. 1988. *Women and the Environment in the Third World: Alliance for the Future.* London: Earthscan.

D'EAUBONNE, F. 1984. *Le Feminism ou la Mort.* Paris: Pierre Horay.

DEERE, C .D. AND LEON DE LEAL, M. 1982. *Women in Andean Agriculture: Peasant Production and Rural Wage Employment in Colombia and Peru.* Geneva: International Labour Organisation.

DEUTSCHE, R. 1991. Boys town. *Environment and Planning D: Society and Space,* 9(1), 5–30.

DIXON, R. 1978. *Rural Women at Work: Strategies for Development in South Asia.* London: Johns Hopkins University Press.

DÖLLING I. 1991. Between hope and helplessness: women in the GDR after the 'Turning Point'. *Feminist Review,* 39, 3–15.

DOMOSH, M. 1991a. Toward a feminist historiography of geography. *Transactions of the Institute of British Geographers,* 16, 95–104.

DOMOSH, M. 1991b. Beyond the frontiers of geographical knowledge. *Transactions of the Institute of British Geographers,* 16, 488–490.

DOWLING, R. 1993. Femininity, place and commodities: a retail case study. *Antipode,* 25, 295–319.

DRIVER, F. 1995 Sub-merged identities: familiar and unfamiliar histories. *Transactions of the Institute of British Geographers,* 20(4), 410–413.

DYCK, I. 1993. Ethnography: a feminist method? *The Canadian Geographer,* 37(1), 52–57.

DYER, R. 1993. *The Matter of Images: Essays on Representations.* London: Routledge.

EINHORN, B. 1991. Where have all the women gone? Women and the women's movement in East Central Europe. *Feminist Review,* 39, 16–36.

ELSHTAIN, J. 1987. *Women and War*. New York: Basic Books.

ELSHTAIN, J. AND TOBIAS, S. 1990. *Women, Militarism and War*. Totowa, NJ: Rowman and Littlefield.

ELSON, D. 1991. *Male Bias in the Development Process*. Manchester: Manchester University Press.

ELSON, D. 1995. *Male Bias in the Development Process*, 2nd edition. Manchester: Manchester University Press.

ELSON, D. AND PEARSON, R. 1981. Nimble fingers make cheap workers: an analysis of women's employment in Third World export manufacturing. *Feminist Review*, 7, 87–107.

ENGLAND, K. 1994. Getting personal: reflexivity, positionality, and feminist research. *The Professional Geographer*, 46(1), 80–89.

ENLOE, C. 1983. *Does Khaki Become You?* Boston: South End Press.

ENNEW, J. 1986. Mujercita y mamacita: girls growing up in Lima. *Bulletin of Latin American Research*, 5(2), 49–66.

ENNEW, J. 1993. Maid of all work. *New Internationalist*, February.

EYLES, J. 1993. Feminist and interpretive method: how different? *The Canadian Geographer*, 37(1), 50–52.

FARRAN, D. 1990. 'Seeking Susan': producing statistical information on young people's leisure, in Stanley, L. (ed.), *Feminist Praxis: Research, Theory and Epistemology in Feminist Sociology*. London: Routledge, pp. 91–102.

FLACSO. 1992. *Mujeres Latinoamericanas en Cifras* (Latin American Women in Statistics). FLACSO (La Facultad Latinoamerica de Ciencias Sociales).

FORBES-MARTIN, S. 1991. *Refugee Women*. London: Zed Press.

FORD, L., DANNER, M. AND YOUNG, G. 1994. Gender inequalities around the world: comparing fifteen nations in five world regions. In Young, G. and Dickerson, B. (eds), *Color, Class and Country: Experiences of Gender*. London: Zed Books.

FORD, S. 1989. Gender, space and the idea of the suburb 1800–1870. Paper presented at the Feminism and Historical Geography one-day conference, London, 11 November.

FOREIGN AND COMMONWEALTH OFFICE. 1995. Outcome of the Fourth World Conference on Women, on hhttp://www.fco.gov.uk/women/index.html, 24/11/95.

FROBEL, F., HEINRICHS, J. AND KREYE, O. 1980. *The New International Division of Labour*. Cambridge: Cambridge University Press.

FUNK, N. 1992. Feminism East and West. In Funk, N. and Mueller, M. (eds), *Gender Politics and Post-Communism*. New York: Routledge, pp. 318–330.

GAINES, J. 1988. White privilege and looking relations: race and gender in feminist film theory. *Screen*, 29(1), 12–27.

GIBSON-GRAHAM, J. K. 1994. 'Stuffed if I know': reflections on post-modern feminist social research. *Gender, Place and Culture*, 1(2), 205–224.

GILBERT, M. 1994. The politics of location: Doing feminist research at 'home', *The Professional Geographer*, 46(1), 90–96.

GOULD, P. 1994. Guest essay/essai sur invitation: sharing a tradition – geographies from the enlightenment. *The Canadian Geographer*, 38(3), 194–200.

GRACE, E. 1990. *Shortcircuiting Labour: Unionising Electronic Workers in Malaysia*. Kuala Lumpur: INSAN.

GREED, C. 1994. *Women and Planning: Creating Gendered Realities*. London: Routledge.

GREGSON, N. 1993. 'The initiative': delimiting or deconstructing social geography? *Progress in Human Geography*, 17(4), 525–530.

GREGSON, N. 1995. And now it's all consumption? *Progress in Human Geography*, 19(1), 135–141.

GREGSON, N. AND LOWE, M. 1994. *Servicing the Middle Classes: Class, Gender and Waged Domestic Labour in Contemporary Britain*. London: Routledge.

GREGSON, N. AND LOWE, M. 1995. 'Home-making': on the spatiality of daily social reproduction in contemporary middle class Britain. *Transactions of the Institute of British Geographers*, 20, 224–235.

GROSZ, E. 1986. What is feminist theory? In Pateman, C. and Grosz, E. (eds), *Feminist Challenges: Social and Political Theory*. Boston: Northeastern University Press, pp. 190–204.

HAFFAJEE, F. 1995. The sisterly republic. *New Internationalist*, March.

HAGGETT, P. 1965. *Locational Analysis in Human Geography*. London: Edward Arnold.

HANSON, S. 1992. Geography and feminism: worlds in collision? *Annals of the Association of American Geographers*, 82, 569–586.

HARAWAY, D. 1991. *Simians, Cyborgs and Women: the Reinvention of Nature*. London: Routledge.

HARDING, S. 1987. Introduction: Is there a feminist method? In Harding, S. (ed.), *Feminism and Methodology*. Bloomington: Indiana University Press, pp. 1–14.

HARRISON, C. 1994. The effects of landscape. In Mitchell, W.J.T., *Landscape and Power*. Chicago: University of Chicago Press, pp. 203–239.

HAYDEN, D. 1981. *The Grand Domestic Revolution: A History of Feminist Designs for American Houses, Neighbourhoods and Cities*. Cambridge, Massachusetts: MIT Press.

HILL COLLINS, P. 1990. *Black Feminist Thought: Knowledge, Consciousness and the Politics of Empowerment*. London: Harper Collins.

HILTY, G. 1993. Georgia O'Keeffe: American and Modern. *Apollo*, 137 (373), March, pp. 194–195.

HODGE, D.C. 1995. Introduction to 'Focus: Should Women Count? The role of quantitative methodology in feminist geographic research'. *The Professional Geographer* 47(4), 426–427.

HOODFAR, H. 1991. Return to the veil: personal strategy and public participation in Egypt. In Redclift, N. and Sinclair, M.T. (eds), *Working Women: International Perspectives on Labour and Gender Ideology*. London: Routledge, pp. 104–124.

HOOKS, B. 1990. *Yearning: Race, Gender and Cultural Politics*. London: Turnaround.

INSTITUTO NACIONAL DE ESTADÍSTICA, GEOGRAFÍA E INFORMÁTICA. 1993. *La Mujer en México: XI Censo General de Población y Vivienda, 1990*. Aguascalientes: INEGI.

INTERNATIONAL LABOUR ORGANISATION. 1995. *Yearbook of Labour Statistics*. Geneva: ILO.

JACKSON, P. 1989. *Maps of Meaning: An Introduction to Cultural Geography*. London: Unwin Hyman.

JACKSON, P. 1994. Black male: advertising and the cultural politics of masculinity. *Gender, Place and Culture*, 1(1), 49–59.

JACKSON, P. AND PENROSE, J. 1993. *Constructions of Race, Place and Nation*. London: UCL Press.

JACOBS, J. 1994. Earth honoring: Western desires and indigenous knowledges. In Blunt, A. and Rose, G. (eds), *Writing Women and Space: Colonial and Postcolonial Geographies*. London: Guildford Press, pp. 169–196.

JARVIE, I.C. 1982. The problem of ethical integrity in participant observation. In Burgess, R.G. (ed.), *Field Research*. London: Allen and Unwin, pp. 68–72.

JAYARATNE, T. 1993. The value of quantitative methodology for feminist research. In Hammersley, M. (ed.), *Social Research: Philosophy, Politics and Practice*. London: Sage, pp. 109–123.

JAYARATNE, T. AND STEWART, A. 1991. Quantitative and qualitative methods in the social sciences. In Fonow, M.M. and Crook, J.A. (eds), *Beyond Methodology: Feminist Scholarship as Lived Research*. Bloomington: Indiana University Press, pp. 85–106.

JOHNSON, L. 1989. Weaving workplaces: sex, race and ethnicity in the Australian textile industry. *Environment and Planning A*, 21(5), 681–684.

JOHNSON, L. AND VALENTINE, G. 1995. Wherever I lay my girlfriend that's my home: the performance and surveillance of lesbian identities in domestic environments. In Bell, D. and Valentine, G. (eds), *Mapping Desire: Geographies of Sexualities*. London: Routledge, pp. 99–113.

JOLY, D. 1992. *Refugees: Asylum in Europe?* London: Minority Rights Publications.

JORDANOVA, L. 1989. *Sexual Visions: Images of Gender in Science and Medicine in the Eighteenth and Twentieth Centuries*. Brighton: Harvester Wheatsheaf.

JULIEN, I. 1994. *The Dark Side of Black* (film). Black Audio/Normal Production for BBC TV. Broadcast April 1994, BBC2 (UK).

KATZ, C. 1994. Playing the field: questions of fieldwork in geography. *The Professional Geographer*, 46(1), 67–72.

KATZ, C. AND MONK, J. (eds), 1993. *Full Circles: Geographies of Women over the Life Course*. London: Routledge.

KEITH, M. 1992. Angry writing: (re)presenting the unethical world of the ethnographer. *Environment and Planning D: Society and Space*, 10(5), 551–568.

KEITH, M. AND PILE, S. (eds), 1993. *Place and the Politics of Identity*. London: Routledge.

KELLY, L. 1988. *Surviving Sexual Violence*. Cambridge: Polity.

KELLY, L. 1990. Journeying in reverse: possibilities and problems in feminist research on sexual violence. In Gelsthorpe, L. and Morris, A. (eds), *Feminist Perspectives in Criminology*. Milton Keynes: Open University Press, pp. 107–114.

KELLY, L., BURTON, S. AND REGAN, L. 1994. Researching women's lives or studying women's oppression? Reflections on what constitutes feminist research. In Maynard, M. and Purvis, J. (eds), *Researching Women's Lives from a Feminist Perspective*. London: Taylor and Francis, pp. 27–48.

KESSLER, S. 1990. The medical construction of gender: case management of intersexed infants. *Signs*, 16(1), 3–26.

KETTEL, B. 1993. New approaches to sustainable development. *Canadian Women's Studies*, 13(3), 11–14.

KINSMAN, P. 1995. Landscape, race and national identity: the photography of Ingrid Pollard. *Area*, 27, 300–310.

KOBAYASHI, A. 1994. Colouring the field: gender, 'race', and the politics of fieldwork. *The Professional Geographer*, 46(1), 73–80.

KOLINSKY, E. 1993. *Women in Contemporary Germany*. Oxford: Berg.

LADNER, J.A. 1987. Introduction to 'Tomorrow's Tomorrow: The Black Woman'. In Harding, S. (ed.), *Feminism and Methodology*. Bloomington: Indiana University Press, pp. 74–83.

LA DUKE, J. AND LUXTON, S. (eds), 1983. *Full Moon: an Anthology of Canadian Women Poets*. Ontario, CA: Quadrat Editions.

LAMB, S. 1994/95. Development, ecology and feminism. *Trouble and Strife*, 29/30, 25–32.

LAURIE, N. 1995. Negotiating Gender: Women and Emergency Employment in Peru. Unpublished PhD thesis. Department of Geography, University College London.

LAWSON, V. 1995. The politics of difference: Examining the quantitative/qualitative dualism in post-structuralist feminist research. *The Professional Geographer*, 47(4), 449–458.

LESLIE, D.A. 1993. Femininity, post-Fordism and the 'new traditionalism'. *Society and Space*, 11(6), 689–708.

LEWIS, C. AND PILE, S. 1996. Woman, body, space: Rio carnival and the politics of performance. *Gender, Place and Culture*, 3(1), 23–42.

LEWIS, J. 1984. The role of female employment in the industrial restructuring and regional development of the UK. *Antipode*, 16, 47–60.

LIM, L. 1991. Women's work in export factories: the politics of a cause. In Tinker, I. (ed.), *Persistent Inequalities: Women and World Development*. New York: Oxford University Press.

LIPIETZ, A. 1997. The post-Fordist world: labour relations, international hierarchy and global ecology. *Review of International Political Economy*, 4(1).

LIPPARD, L. 1994. Passenger on the shadows. In *Aperture* 137 Brush fires in the social landscape – David Wojnarowicz: 6–25.

LIVINGSTONE, D. 1986. Nature. In Jonhston, R., Gregory, D. and Smith, D.M. (eds), *The Dictionary of Human Geography*, second edition. Oxford: Basil Blackwell, pp. 314–315.

LIVINGSTONE, D. 1992. *The Geographical Tradition*. Oxford: Basil Blackwell.

LLOYD, G. 1984. *The Man of Reason: 'Male' and 'Female' in Western Philosophy*. London: Methuen.

LYNES, B.B. 1989. *O'Keeffe, Stieglitz and the Critics, 1916–1929*. Ann Arbor and London, UMI Research Press.

MACKENZIE, S. AND ROSE, D. 1983. Industrial change, the domestic economy and home life. In Anderson, J., Duncan, S. and Hudson, R. (eds), *Redundant Spaces in Cities and Regions?* London: Academic Press, pp. 155–200.

MACKENZIE, S., FOORD, J. AND McDOWELL, L. 1980. Women's place – women's space. *Area*, 12(1), 47–51.

MADGE, C. 1994. The ethics of research in the 'Third World'. In Robson, E. and Willis, K. (eds), *DARG Monograph No. 8: Postgraduate Fieldwork in Developing Areas* (Developing Areas Research Group of the Institute of British Geographers), pp. 91–102.

MAGUIRE, P. 1987. *Doing Participatory Research: a Feminist Approach*. Amherst: University of Massachusetts, Center for International Education.

MASSEY, D. 1991. Flexible sexism. *Environment and Planning D: Society and Space*, 9, 31–57.

MASSEY, D. 1994. *Space, Place and Gender*. Cambridge: Polity.

MASSINGHAM, B. 1982. *A Century of Gardeners*. London: Faber & Faber.

MATRIX. 1984. *Making Space: Women and the Man-Made Environment*. London: Pluto Press.

MATTINGLY, D. J. and FALCONER-AL-HINDI, K. 1995. Should women count? A context for the debate. *The Professional Geographer*, 47(4), 427–437.

MAYNARD, M. 1994. Methods, practice and epistemology: The debate about feminism and research. In Maynard, M. and Purvis, J. (eds) *Researching Women's Lives from a Feminist Perspective*. London: Taylor and Francis, pp. 10–26.

MAYNARD, M. and PURVIS, J. (eds) 1994. *Researching Women's Lives from a Feminist Perspective*. London: Taylor and Francis.

McDOWELL, L. 1983. Towards an understanding of the gender division of urban space. *Environment and Planning D: Society and Space*, 1(1), 59–72.

McDOWELL, L. 1989. Gender divisions. In Hamnett, C., McDowell, L. and Sarre, P. (eds), *The Changing Social Structure*. London: Sage, pp. 158–198.

MCDOWELL, L. 1991. Life without father and Ford: the new gender order of post-Fordism. *Transactions of the Institute of British Geographers*, 16(4), 400–419.

MCDOWELL, L. 1992. Doing gender: Feminism, feminists and research methods in human geography. *Transactions of the Institute of British Geographers*, 17(4), 399–416.

MCDOWELL, L. 1993a. Space, place and gender relations. Part I: Feminist empiricism and the geography of social relations. *Progress in Human Geography*, 17(2), 159–179.

MCDOWELL, L. 1993b. Space, place and gender relations. Part II: Identity, difference and feminist geometries and geographies. *Progress in Human Geography*, 17(3), 305–318.

MCDOWELL, L. 1995. Body work: heterosexual gender performances in city workplaces. In Bell, D. and Valentine, G. (eds), *Mapping Desire: Geographies of Sexualities*. London: Routledge, pp. 75–95.

MCDOWELL, L. and COURT, G. 1994. Missing subjects: gender, sexuality and power in merchant banking. *Economic Geography*, 70, 229–251.

MCDOWELL, L. and MASSEY, D. 1984. A woman's place. In Massey, D. and Allen, J. (eds), *Geography Matters*. Cambridge: Cambridge University Press, pp. 128–147.

MCLAFFERTY, S. L. 1995. Counting for women. *The Professional Geographer*, 47(4), 436–442.

MERCHANT, C. 1981. *The Death of Nature: Women, Ecology and the Scientific Revolution*. London: Wildwood House.

MERNISSI, F. 1987. *Beyond the Veil: Male–Female Dynamics in Modern Muslim Society*. Bloomington: Indiana University Press.

MIES, M. 1983. Towards a methodology for feminist research. In Bowles, G. and Klein, R. D. (eds) *Theories of Women's Studies*. London: Routledge and Kegan Paul, pp. 117–139.

MIES, M. 1991. Women's research or feminist research? The debate surrounding feminist science and methodology. In Fonow, M. M. and Cook, J. A. (eds), *Beyond Methodology: Feminist Scholarship as Lived Research*. Bloomington: Indiana University Press, pp. 60–84.

MILLER, J. 1985. *You Can't Kill the Spirit: Women in a Welsh Mining Valley*. London: The Women's Press.

MILLER, R. 1983. The Hoover® in the garden: middle class women and suburbanisation, 1850–1920. *Environment and Planning D: Society and Space*, 1, 73–87.

MILLER, R. 1991. Selling Mrs Consumer: advertising and the creation of suburban socio-spatial relations, 1910–1930. *Antipode*, 23(3), 263–301.

MITTER, S. 1986. *Common Fate, Common Bond. Women in the Global Economy*. London: Pluto Press.

MOHANTY, C. T. 1991. Under Western eyes: feminist scholarship and colonial discourse. In Mohanty, C. T., Russo, A. and Torres, L. (eds), *Third World Women and the Politics of Feminism*. Bloomington: Indiana University Press, pp. 51–80.

MOMSEN, J. 1992. *Women and Development in the Third World*. London: Longman.

MOMSEN, J. H. and KINNAIRD, V. (eds) 1993. *Different Places, Different Voices: Gender and Development in Africa, Asia and Latin America*. London: Routledge.

MOMSEN, J. and TOWNSEND, J. (eds) 1987. *Geography of Gender in the Third World*. London: Hutchinson.

MONK, J. (1992) Gender in the landscape: expressions of power and meaning. In Anderson, K. and Gayle, F. (eds), *Inventing Places*. Sydney: Longman Cheshire, pp.123–138

MONK, J. and HANSON, S. 1982. On not excluding half of the human in human geography. *The Professional Geographer*, 34(1), 11–23.

MONK, J. and NORWOOD, V. 1987. Introduction: perspectives on gender and landscape. In Monk, J. and Norwood, V. (eds), *The Desert is No Lady: Southwestern Landscapes in Women's Writings and Art*. New Haven: Yale University Press, pp.1–9.

MOORE, H. 1994. *A Passion for Difference*. Cambridge: Polity.

MORRIS, M. S. 1994. Home and Garden: Landscape and Gender in English Culture 1880–1930. Unpublished PhD thesis, Department of Geography, University of Nottingham.

MORRIS, M. S. 1996. 'Tha'lt be like a blush-rose when tha' grows up, my little lass': English cultural and gendered identity in *The Secret Garden*. *Environment and Planning D: Society and Space*, 14, 59–78.

MOSER, C. 1993. Adjustment from below: low-income women, time and the triple role in Guayaquil, Ecuador. In Radcliffe, S. and Westwood, S. (eds), *'Viva' Women and Popular Protest in Latin America*. London: Routledge.

MOSS, P. 1993. Focus: feminism as method. *The Canadian Geographer*, 37(1), 48–49.

MOSS, P 1995. Embeddedness in practice, numbers in context: the politics of knowing and doing. *The Professional Geographer*, 47(4), 442–449.

NASH, C. 1994. Remapping the body/land: new cartographies of identity, gender and landscape in Ireland. In Blunt, A. and Rose, G. (eds), *Writing Women and Space: Colonial and Postcolonial Geographies*. London: Guilford Press.

NASH, C. 1996. Reclaiming vision: looking at landscape and the body. *Gender, Place and Culture*, 3(2), 149–169.

NAST, H. 1994. Opening remarks on 'In the field'. *The Professional Geographer*, 46(1) 54–66.

NESMITH, C. and RADCLIFFE, S. 1993. Remapping Mother Earth: a geographical perspective on environmental feminisms. *Environment and Plannning D: Society and Space*, 11, 379–394.

OAKLEY, A. 1990. Interviewing women: a contradiction in terms. In Roberts, H. (ed.), *Doing Feminist Research*. London: Routledge, pp. 30–61. (Originally published in 1981 by Routledge and Kegan Paul.)

OLDENBURG, V. T. 1990. Lifestyle as resistance: the case of the courtesans of Lucknow, India. *Feminist Studies*, 16(2), 259–287.

ONG, A. 1987. *Spirits of Resistance and Capitalist Discipline: Factory Women in Malaysia*. Albany: State University of New York Press.

OPEN UNIVERSITY. 1991. *Public Space Public Work*. Television programme broadcast as part of *Issues in Women's Studies*, U207. Milton Keynes: Open University.

OPIE, A. 1992. Qualitative research, appropriation of the 'other' and empowerment. *Feminist Review*, 40(1), 52–69.

ORTNER, S. 1974. Is Female to Male and Nature to Culture? In Rosaldo, M. and Lamphere, L. (eds) *Women, Culture and Society*. Stanford: Stanford University Press, pp. 67–88

PAIN, R. (1991) Space, sexual violence and social control: integrating geographical and feminist analyses of women's fear of crime. *Progress in Human Geography*, 15(4), 415–431.

PATAI, D. 1991. US academics and Third World women: is ethical research possible? In Gluck, S. B. and Patai, D. (eds), *Women's Words: the Feminist Practice of Oral History*. London: Routledge, pp. 137–154.

PEAKE, L. 1993. Race and sexuality: challenging the patriarchal structuring of urban social space. *Environment and Planning D: Society and Space*, 11, 415–432.

PEAKE, L. 1994. Proper words in proper places ... or of young turks and old turkeys. *The Canadian Geographer*, 38(3), 204–206.

PEARSON, R. 1986. Latin American women and the New International Division of Labour: a reassessment. *Bulletin of Latin American Research*, 5(2), 67–79.

PENROSE, J., BONDI, L., MCDOWELL, L., KOFMAN, E., ROSE, G. and WHATMORE, S. 1992. Feminists and feminism in the academy. *Antipode*, 24(3), 218–237.

PETERSON, V. S. and RUNYAN, A. 1993. *Global Gender Issues*. Oxford: Westview Press.

PILE, S. 1991. Practising interpretative geography. *Transactions of the Institute of British Geographers*, 16(4), 458–469.

PINNEGAR, S. 1996. Are our young turks simply trendy boys? *Women and Geography Study Group Newsletter*, 1, 20–26.

PLUMWOOD, V. 1992. Feminism and ecofeminism: beyond the dualistic assumption of women, men and nature. *The Ecologist*, 22(1), Jan/Feb, 8–13.

POLLARD, I. 1993. Another view. *Feminist Review*, 45, 46–50.

PRATT, A. 1992. Review of 'Laura Ashley: a life by design' by A. Sebba. *Journal of Rural Studies*, 8(1), 126–127.

PRATT, G. 1992. Spatial metaphors and speaking positions. *Environment and Planning D: Society and Space*, 10, 241–244.

PRATT, G. and HANSON, S. 1991. On the links between home and work: family-household strategies in a buoyant labour market. *International Journal of Urban and Regional Research*, 15, 55–74.

PRATT, G. and HANSON, S. 1993. Women and work across the life course: moving beyond essentialism. In Katz, C. and Monk, J. (eds), *Full Circles: Geographies of Women over the Life Course*. London: Routledge.

PRATT, G. and HANSON, S. 1994. Geography and the construction of difference. *Gender, Place and Culture*, 1(1), 5–31.

PRATT, G. and HANSON, S. 1995. *Gender, Work and Space*. London: Routledge.

PRATT, M. L. 1992. *Imperial Eyes: Travel Writing and Transculturation*. London: Routledge.

PRESTON, V., MCLAFFERTY, S. and HAMILTON, E. 1993. The impact of family status on black, white and Hispanic women's commuting. *Urban Geography*, 14, 228–250.

PUGH, A. 1990. My statistics and feminism – a true story. In Stanley, L. (ed.), *Feminist Praxis: Research, Theory and Epistemology in Feminist Sociology*. London: Routledge, pp. 103–112.

RADCLIFFE, S. 1990. Ethnicity, patriarchy, and incorporation into the nation: female migrants as domestic servants in Peru. *Environment and Planning D: Society and Space*, 8, 379–398.

RADCLIFFE, S. 1994. (Representing) post-colonial women: authority, difference and feminisms. *Area*, 26(1), 25–32.

RADCLIFFE, S. 1996. Gendered nations: nostalgia, development and territory in Ecuador. *Gender, Place and Culture*, 3(1), 5–22.

REINHARZ, S. 1992. *Feminist Methods in Social Research*. Oxford: Oxford University Press.

REUTHER, R. 1975. *New Woman, New Earth*. New York: Seabury Press.

REYNOLDS, D. 1994. Political geography: closer encounters with the state, contemporary political economy and social theory. *Progress in Human Geography*, 18(2), 234–247.

ROBERTS, M. 1991. *Living in a Man-Made World: Gender Assumptions in Modern Housing Design*. London: Routledge.

ROCHELEAU, D. 1995. Maps, numbers, text, and context: mixing methods in feminist political ecology. *The Professional Geographer*, 47(4), 458–466.

RODDA, A. 1991. *Women and the Environment*. London: Zed Books.

ROHNSTOCK, K. (ed.) 1991. *Handbuch Wegweiser für Frauen in den fünf neuen Bundesländern* (Handbook Guide for Women in the Five New Federal German States). Berlin: Frauenbuch Basisdruck.

ROSALDO, M. Z. 1974. Woman, culture, society: a theoretical overview. In Rosaldo, M. Z. and Lamphere, L. (eds), *Woman, Culture, Society*. Stanford: Stanford University Press, pp. 17–42.

ROSE, D. 1984. Rethinking gentrification: beyond the uneven development of marxist theory. *Environment and Planning D: Society and Space*, 2, 47–74.

ROSE, D. 1993. On feminism, method and methods in human geography: an idiosyncratic overview. *The Canadian Geographer*, 37(1), 57–61.

ROSE, G. 1988. Locality, politics and culture: Poplar in the 1920s. *Environment and Planning D: Society and Space*, 6, 151–168.

ROSE, G. 1993. *Feminism and Geography: The Limits of Geographical Knowledge*. Cambridge: Polity.

ROSE, G. 1995a. Distance, surface, elsewhere: a feminist critique of the space of phallocentric self/knowledge. *Environment and Planning D: Society and Space*, 13(6), 761–781.

ROSE, G. 1995b. Tradition and paternity: same difference? *Transactions of the Institute of British Geographers*, 20, 414–416.

ROSENBLUM, N. 1992. Georgia O'Keeffe (book review). *Art Journal*, 51(1), Spring, 105–113.

ROSENFELD, P. 1921. From 'American painting', *Dial*, 71, December, pp. 666–670, quoted in Lynes, B. B., 1989, *O'Keeffe, Stieglitz and the Critics, 1916–1929*, Ann Arbor and London: UMI Research Press, pp.171–172.

SAID, E. W. 1993. *Culture and Imperialism*. London: Vintage (1994 edition), pp. 95–116.

SCHOENBERGER, E. 1991. The corporate interview as an evidentiary strategy in economic geography. *The Professional Geographer*, 43, 180–189.

SCHOENBERGER, E. 1992. Self-criticism and self-awareness in research: a reply to Linda McDowell. *The Professional Geographer*, 44(2) 215–218.

SCOTT, A. 1991. Informal sector or female sector?: gender bias in urban labour market models. In Elson, D. (ed.), *Male Bias in the Development Process*. Manchester: Manchester University Press.

SCOTT, J. W. 1992. 'Experience'. In Butler, J. and Scott, J. W. (eds), *Feminists Theorize the Political*. London: Routledge, pp. 22–40.

SEAGER, J. 1993. *Earth Follies: Feminism, Politics and the Environment*. London: Earthscan.

SEAGER, J. and OLSON, A. 1986. *Women in the World: An International Atlas*. London: Pan.

SEBBA, A. 1990. *Laura Ashley: a Life by Design*. London: Weidenfeld and Nicolson.

SEYMOUR, S., DANIELS, S. and WATKINS, C. 1994. *Estate and empire: Sir George Cornewall's management of Moccas, Herefordshire and La Taste, Grenada, 1771–1819*. Working Paper 28. Nottingham: University of Nottingham, Department of Geography.

SHAFFIR, W. and STEBBINS, R. (eds) 1991. *Experiencing Fieldwork. An Inside View of Qualitative Research*. London: Sage.

SHIVA, V. 1988. *Staying Alive: Women, Ecology and Survival*. London: Zed Books.

SHIVA, V. and MIES, M. 1993. *Ecofeminism*. London: Zed Books.

SILK, J. 1995. Time rolling forward, politics rolling back? *WGSG (of RGS-IBG) Newsletter*, 3, 19–20.

SKELTON, T. 1995a. Boom, by bye: Jamaican ragga and gay resistance. In Bell, D. and Valentine, G. (eds) *Mapping Desire: Geographies of Sexualities*. London: Routledge. pp. 264–283.

SKELTON, T. 1995b. 'I sing dirty reality, I am out there for the ladies', Lady Saw: women and Jamaican ragga music, resisting reality. *Phoebe: Journal of Feminist Scholarship, Theory and Aesthetics*, 7(1/2), 86–104.

SMITH, N. 1993. Homeless/global: scaling places. In Bird, J. *et al.* (eds), *Mapping the Futures: Local Cultures, Global Change*. London: Routledge, pp. 87–119.

SMITH, N. and KATZ, C. 1993. Grounding metaphor: towards a spatialized politics. In Keith, M. and Pile, S. (eds), *Place and the Politics of Identity*. London: Routledge, pp. 67–83.

SMITH, S. 1994. Soundscape. *Area*, 26(3), 232–240.

SOJA, E. 1985. The spatiality of social life: towards a transformative theory. In Gregory, D. and Urry, J. (eds), *Social Relations and Spatial Structures*. London: Macmillan.

SOJA, E. 1989. *Postmodern Geographies*. London: Verso.

STACEY, J. 1988. Can there be a feminist ethnography? *Women's Studies International Forum*, 11(1), 21–27.

STAEHELI, L. and LAWSON, V. 1994. A discussion of 'women in the field': the politics of feminist fieldwork. *The Professional Geographer*, 46(1), 96–102.

STANDING, H. 1991. *Dependence and Autonomy: Women's Employment and the Family in Calcutta*. London: Routledge.

STANLEY, L. and WISE, S. 1993. *Breaking Out Again: Feminist Ontology and Epistemology*. London: Routledge.

STODDART, D. 1986. *On Geography and its History*. Oxford: Basil Blackwell.

STODDART, D. R. 1991. Do we need a feminist historiography of geography – and if we do, what should it be? *Transactions of the Institute of British Geographers*, 16, 484–487.

TAYLOR, P. 1994. The state as a container: territoriality in the modern world-system. *Progress in Human Geography*, 18(2), 151–162.

TIVERS, J. 1978. How the other half lives: the geographical study of women. *Area*, 10(4), 302–306.

TIVERS, J. 1985. *Women Attached: The Daily Lives of Women with Young Children*. London: Croom Helm.

TÓTH, O. 1992. No envy, no pity. In Funk, N. and Mueller, M. (eds), *Gender Politics and Post-Communism*. New York: Routledge, pp. 213–223.

TOWNSEND, J. in collaboration with Ursula Arrevillaga, Jennie Bain, Socorro Cancino, Susan F. Frenk, Silvana Pacheco and Elia Perez. 1995. *Women's Voices from the Rainforest*. London: Routledge.

TUAN, Y.-F. 1974. *Topophilia: a Study of Environment Perception, Attitudes and Values*. Englewood Cliffs, NJ: Prentice Hall.

UNHCR. 1989. *Refugee Women: a Selected and Annotated Bibliography*. Geneva: UNHCR.

United Nations Conference on Environment and Development (UNCED). 1992. *Earth Summit '92*. London: Regency Press.

VALENTINE, G. 1989. The geography of women's fear. *Area*, 21, 385–390.

VALENTINE, G. 1992. Images of danger: women's sources of information about the spatial distribution of male violence. *Area*, 24(1), 22–29.

VALENTINE, G. 1993a. (Hetero)sexing space: lesbian perceptions and experiences in everyday spaces. *Environment and Planning D: Society and Space*, 11, 395–413.

VALENTINE, G. 1993b. Desperately seeking Susan: a geography of lesbian friendships. *Area*, 25(2), 109–116.

VALENTINE, G. 1993c. Negotiating and managing multiple sexual identities: lesbian time–space strategies. *Transactions of the Institute of British Geographers*, 18(2), 237–248.

VARGAS, V. 1991. The women's movement in Peru: streams, spaces and knots. *European Review of Latin American and Caribbean Studies*, 50, 7–47.

WARE, V. 1992. *Beyond the Pale: White Women, Racism and History.* London: Verso.

WARREN, K. J. (ed.) 1994. *Ecological Feminism.* London: Routledge.

WEKERLE, G. R. and WHITZMAN, C. 1995. *Safe Cities: Guidelines for Planning, Design and Management.* London: Van Nostrand Reinhold.

WILLIAMS, J. 1994. Singing the Blues: a Feminist Analysis of Post-Natal Depression. Unpublished MA Thesis, Edge Hill College, Ormskirk, UK.

WILLIAMS, S. 1994. *The Oxfam Gender Training Manual.* Oxford: Oxfam.

WILLIS, K. 1995. Women's social networks as a household resource: evidence from Southern Mexico. Unpublished paper for Latin America Studies Seminar, Liverpool University, November.

WILSON, F. 1993. Workshops as domestic domains: reflections on small-scale industry in Mexico. *World Development*, 21(6), 67–80.

Women and Geography Study Group of the Institute of British Geographers. 1984. *Geography and Gender: an Introduction to Feminist Geography.* London: Hutchinson and Explorations in Feminism Collective.

ZELINSKY, W., MONK, J. and HANSON, S. 1982. Women and geography: a review and prospectus. *Progress in Human Geography*, 6(3), 317–366.

Index